Treatise on Intuitionistic Type Theory

LOGIC, EPISTEMOLOGY, AND THE UNITY OF SCIENCE

VOLUME 22

Logic, Epistemology, and the Unity of Science aims to reconsider the question of the unity of science in light of recent developments in logic. At present, no single logical, semantical or methodological framework dominates the philosophy of science. However, the editors of this series believe that formal techniques like, for example, independence friendly logic, dialogical logics, multimodal logics, game theoretic semantics and linear logics, have the potential to cast new light on basic issues in the discussion of the unity of science.

This series provides a venue where philosophers and logicians can apply specific technical insights to fundamental philosophical problems. While the series is open to a wide variety of perspectives, including the study and analysis of argumentation and the critical discussion of the relationship between logic and the philosophy of science, the aim is to provide an integrated picture of the scientific enterprise in all its diversity.

For further volumes:
http://www.springer.com/series/6936

Johan Georg Granström

Treatise on Intuitionistic Type Theory

Springer

Johan Georg Granström
Seestrasse 49
CH-8810 Horgen
Switzerland
georg.granstrom@gmail.com

ISBN 978-94-007-1735-0 e-ISBN 978-94-007-1736-7
DOI 10.1007/978-94-007-1736-7
Springer Dordrecht Heidelberg London New York

Library of Congress Control Number: 2011929986

Printed on acid-free paper

Springer is part of Springer Science+Business Media (www.springer.com)

Contents

List of Figures

List of Tables

Omnia cum veterum sint explorata libellis,
Multa loqui breviter sit novitatis opus.[†]

[†]Preface to Phocas' grammar, quoted in de Bury, *Philobiblon*, p. 50.

Introduction

THE PREREQUISITES for understanding this book have been kept at a minimum. All that is required is some background in mathematics and the ability to follow a rigorous proof. Background in philosophy and computer science is helpful but not necessary. This book is about intuitionistic type theory, but no background in logic is required, even though some familiarity with natural deduction is an advantage. In fact, profound knowledge of extensional set theory can be an impediment, rather than a help, since it is difficult to forget what one already knows. The easiest way to learn intuitionistic type theory is to disregard any preconceptions about logic and set theory and start afresh with the definitions and axioms of intuitionistic type theory. Only after having understood the whole system and its methodology, one should make a comparison with what one knew before.

The kind of type theory presented in this book has been variously called *intuitionistic* type theory, *constructive* type theory, *dependent* type theory, and, after its first expounder, Martin-Löf's type theory. As often is the case when one subject has many different names, there are different nuances to them—so also in this case. I have chosen the name intuitionistic type theory because it was the first name applied to the subject and because the type theory in question is the natural adaption of earlier type theories, e.g., the ramified and simple type theories, to *intuitionistic* principles.

In this book I will present intuitionistic type theory together with my own contributions to it, resulting in a version of intuitionistic type theory which is essentially backwards compatible with Martin-Löf's version.[1] My technical contributions are the following:

(1) coinductive definitions of sets are treated on a par with inductive definitions (cf. Def. 4 on p. 64);

(2) consequently, the set of functions from one set to another is coinductively defined (cf. p. 99);

(3) the introduction of separate forms of assertion for canonical and non-canonical sets and elements (cf. p. 64 and p. 91);

(4) the introduction of computation rules into the language of intuitionistic type theory (cf. p. 91);

[1]Martin-Löf's 1984 book *Intuitionistic Type Theory* contains the essential ideas of intuitionistic type theory, but several important contributions to the system have been presented only in the form of lectures. At present, the most authoritative presentation of intuitionistic type theory is Nordström, Petersson and Smith, *Programming in Martin-Löf's Type Theory*.

(5) the default way of computing type-theoretic terms is now eager, as opposed to lazy (cf. p. 87), though lazy evaluation is still an option;

(6) in my version of the substitution calculus, weakening is explicit, and the freshness condition on variables can be removed (cf. p. 129).

In addition to the above, I have made some minor changes in syntax and presentation. The development of intuitionistic type theory with these modifications forms a kind of technical core to this book, made up by Chapters III to V. Chapters I and II contain philosophical background which I found necessary for my own understanding of intuitionistic type theory, and I hope that they are relevant to a wider audience. Some final thoughts on intuitionism and its adaption are presented in Chapter VI.

In this book, a plenitude of topics are touched upon, many of them only *en passant.* Therefore, I have provided an index of subjects at the end: it is in general easier to find a specific topic through the index than through the table of contents, since not all topics have a perfectly logical place in the order of presentation.

Chapters I, II, and VI also present a coherent epistemology: this epistemology is my version of Martin-Löf's intuitionistic epistemology.[2] At some points, this epistemology is, as it were, unbrindled, and excursions are made into the philosophy of science: most notably, in Chapter I, Section 4, on the lingua characteristica, and in Chapter VI, Section 2, on the law of excluded middle. But, clearly, much more could be said about the relavance of the intuitionistic epistemology for general science.

Most of the topics treated of lie in the intersection of at least two of three subjects: philosophy, mathematical logic, and computer science. This is in part due to the nature of the subject matter and in part due to my own preferences and background. Thus, the present book is, in the wide sense of the word, an *interdisciplinary* work. Two problems with any interdisciplinary work are that the style of presentation has to be chosen from *one* of the subjects, and that the author cannot equally be an expert in all fields. Here I have chosen the style of presentation of a work in philosophy, because it can be used to express thoughts in almost any domain of discourse, whereas, e.g., the style of presentation usually employed in computer science (the system of referencing, etc.) is inappropriate for philosophical work—to my mind, it would, for example, be rude to refer to Aristotle by a number in brackets.

My first interest in intuitionistic type theory was motivated by my interest in computer science and the search for the perfect programming language. Unfortunately I have not found the time to write as much about the computer scientific aspects of intuitionistic type theory as I would have liked. The litterature in this field is vast, and technicalities often obscure philosophically relevant points.

---

[2]Presented in a series of articles: Martin-Löf, 'Truth of a proposition, evidence of a judgement, validity of a proof', 'A Path from Logic to Metaphysics', 'Analytic and synthetic judgements in type theory', 'Verificationism Then and Now', 'On the meanings of the logical constants and the justifications of the logical laws'. The pedigree of intuitionistic epistemology include philosophers such as Frege, Husserl, Brouwer, and Becker.

Looking at the citations and references that I bring in to support the doctrines of intuitionistic type theory, it seems as if all logicians and philosophers of all times support them. This is of course not the case: many philosophers have said a lot of things. The quotations are carefully chosen to support a particular point of view on a particular topic. I have chosen this approach because I want to present the whole of intuitionistic type theory uninterrupted, so to say. The critical discussion is saved until Chapter VI, Section 3, in which I try to show that the two competitive approaches to the foundations of mathematics, viz., formalism and set-theoretical Platonism, are ultimately untenable. Thus, I have tried to avoid being pugnacious except in this section, though, admittedly, there are some scattered apologetic remarks elsewhere too.

CHAPTER I

Prolegomena

IT IS FITTING to begin this book on intuitionistic type theory by putting the subject matter into perspective. The purpose of this chapter is to relate intuitionistic type theory to the old dream of a *lingua characteristica*. The line of thought which leads to the *lingua characteristica* can be briefly summarized as follows. Man thinks about things and expresses his thoughts in words: this leads to the threefold correspondence discussed in the first section. Based on this correspondence, the acts of the mind, or the thoughts themselves, are analysed and divided in the second section. The third section treats of a theory of meaning according to which the complex is understood by synthesizing, as it were, the meanings of its parts. The fourth and last section of this chapter gives an account of the history of the *lingua characteristica*.

§ 1. A threefold correspondence

The word logic, or rather its Early English spelling logike, is a direct transliteration of the word λογική, first used in its present sense by Zeno the Stoic.[1] The word λογική is in turn derived from the word λόγος with a wide range of meanings from the concrete, word or speech, to the abstract, discourse or reason.[2] *Prima facie* and according to the opinion of most ancient philosophers, concepts are derived from things and words are expressions of thoughts.[3] The classical view on this threefold correspondence is that things have priority over thoughts, and thoughts over words, as eloquently expressed by Cajetan in the beginning of his commentary of Aristotle's *Categories*:[4]

"And even if we have to maintain this interpretation of the intention of this book, we must not forget what Avicenna so aptly says at the beginning of his Logic, namely, that to treat of words does not pertain to logical discussions on purpose, but it is only a sort of necessity that forces this on us, because the things so conceived we cannot express, teach, unite, and arrange, but by the help of words. For if we were able to carry out all these things without the use of external words, satisfied by the use of internal speech alone, or if by other signs would these things be achieved, it would be pointless to treat of words. So

[1]Cf. Diogenes, *Lives of Eminent Philosophers*, Ch. 7, in particular n. 32, sqq.; and Cicero, *De Fato*, n. 1.

[2]From *oratio* to *ratio*, to use two common Latin translations of the word λόγος.

[3]Cf. Aristotle, *Perih.*, Ch. 1.

[4]To aid the understanding of the first part of this quotation, it should be added that Cajetan's interpretation of Aristotle's point of view is that words are signs of concepts and that concepts are signs of things (cf. ibid., Ch. 1, 16a4).

J.G. Granström, *Treatise on Intuitionistic Type Theory*, Logic, Epistemology, and the Unity of Science 22, DOI 10.1007/978-94-007-1736-7_1,

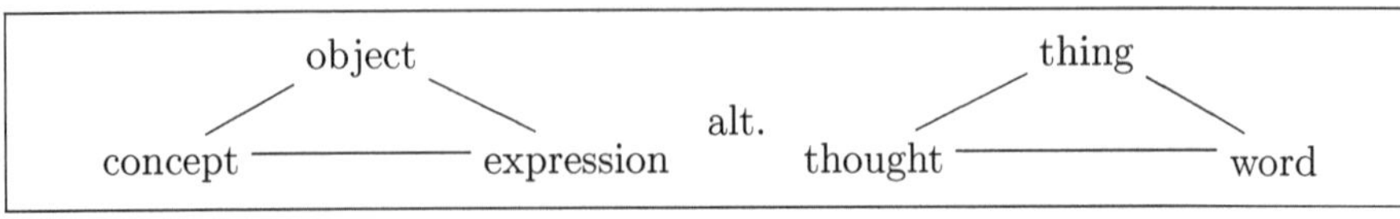

Figure 1. The relation between object, concept, and expression, as well as between thing, thought, and word.

if one were to ask whether it is words or things which are principally treated of here, we have to say that it is things, though not absolutely, but insofar as they are conceived in an incomplex manner, and, by consequent necessity, insofar as signified by words."[5]

With Aristotle and Cajetan, I will defend the thesis that thought has a kind of priority over words and that the concept has a kind of priority over its expression.[6] Cajetan also brings up another important point in the above quotation, namely, that it is a sort of necessity which forces the treatment of words upon us because, if a thought is to be communicated, then there has to be words for it. This insight is a kind of contrapositive to Wittgenstein's famous dictum that "Whereof one cannot speak, thereof one must be silent."[7]

Let us consider the relation between words, thoughts, and things in greater detail. As a general rule, the more experienced we are in a particular field the less we pay attention to the signs and expressions of the field and even to their meanings; instead our attention is entirely focused on the things. As an example, consider the driver of an automobile approaching a stop sign. The experienced driver does not pay attention to the word stop, nor to the red colour or to the hexagonal shape; perhaps he does not even become conscious of the significance of the sign—he simply stops, habitually, as it were. In a like manner, the scientist learns to see through the expressions of his field and, to a certain extent, even their meanings.[8] This is all well, except in philosophy, logic, and related subjects, where we *have to see* the words and their meanings to be able to investigate them. Note, however, that the meaning of an expression never should be considered an object in its own right, as this would blur the distinctions made in this chapter.

The names given to the three terms of this correspondence differ between authors: De Morgan, for example, writes object, idea, and name; Peirce writes object, mind, and sign;[9] Frege has, in Geach's translation, object and concept as two of the terms;[10] and the title of Martin-Löf's lecture *Categories of expres-*

[5]Cajetan, *In Praed.*, Ch. 1.

[6]Cf. Ch. VI, § 3 of this book.

[7]Wittgenstein, *Tractatus*, § 7: "Wovon man nicht sprechen kann, darüber muss man schweigen." The exact contrapositive of Cajetan's point is that, if there are no words for a thought, then it cannot be communicated. I take this to be tantamount to Wittgenstein's dictum.

[8]Another example, due to Descartes, is that it may happen that we remember something that somebody told us, without remembering in which language it was spoken ('The World or Treatise on Light', Ch. 1, n. 4).

[9]De Morgan, *Formal Logic*, Ch. 2; Peirce, 'On the Algebra of Logic: a contribution to the philosophy of notation', Ch. 1.

[10]'On Concept and object' is Geach's translation of the title of Frege's article 'Über Begriff und Gegenstand'.

sions, meanings or objects? suggests yet another terminology.[11] Some choice has to be made. As can be seen from Figure 1, I have adopted the triple object, concept, and expression; sometimes I also use the more old-fashioned triple thing, thought, and word. It remains to provide some justification for this choice.

It is clear that *object* is a better choice than *thing*, because *thing* brings the thoughts to sensible things. Numbers, for example, are called objects, but seldom called things.[12] Thus the first term of the correspondence is object.

For the second term, De Morgan has *idea*, Peirce has *mind*, Martin-Löf has *meaning*, and a fourth suggestion is *concept*. As opposed to De Morgan's term *idea*, the term *concept* has the advantage of being more exact. Idea can be understood in a fitting way, but it also has several other possible interpretations. Peirce's suggestion *mind*, by which he evidently means *something in the mind*, is ruled out for the same reason: there are other things in the mind except concepts. Martin-Löf's term, *meaning*, is ambiguous between the relation between word and thought, and the thoughts themselves. Even though this ambiguity sometimes is an advantage, I have chosen the word *concept* as the second term.[13]

As for the third term of the correspondence, the suggestions are *word*, *name*, *sign*, and *expression*. As seen below, *word* is too general because words need not express anything. On the other hand, *name* is too specific since it is usually taken to exclude complex expressions. Recall Aristotle's well-known definition of a name, or noun, ὄνομα in Greek, as

"a sound having meaning established by convention alone but no reference whatever to time, while no part of it has any meaning, considered apart from the whole."[14]

If we disregard the distinction between written and spoken language, we may well read *word* for *sound*; moreover, if we want to include complex expressions, the last part of the definition has to be removed. We end up defining an expression as words having meaning established by convention alone but no reference whatever to time. Sometimes even this definition is too restrictive and any meaningful combination of words will be regarded as an expression.

Peirce, the founder of modern semiotic, stresses the importance of signs in the study of logic.[15] I agree with this emphasis, but have still ruled out *sign* as the third term of the correspondence. Consider Arnauld's explanation of the relation between sign and object:

"When one considers an object in itself and according to its own being, then he thinks of that object simply as a thing; but when he considers an object as representing some other object, then the first object is being thought

[11]This lecture was held in Tampere in May 2007 and in Stockholm in November the same year.

[12]Cf. Maritain, *The Degrees of Knowledge*, Ch. 3, § 10.

[13]The identification of concept and meaning is rejected by Husserl, *Log. Unt. II*, Pt. 1, Inv. 1, § 33, but the distinction is subtle and he subsequently often speaks as if they were identified. Cf. Maritain, *The Degrees of Knowledge*, Ch. 3, § 24.

[14]Aristotle, *Perih.*, Ch. 2.

[15]E.g., Peirce, 'On the Algebra of Logic: a contribution to the philosophy of notation', pp. 180–181.

of as a sign. Maps and pictures are ordinarily regarded as signs. Thus, when we consider an object as sign, we consider two things: the sign as thing, and the thing signified by the sign. It is of the nature of a sign that the idea of the sign excites the idea of the thing signified by the sign."[16]

Arnauld continues by explaining the traditional division of signs into natural and conventional signs, where the latter need not have any natural connection with the thing signified, and states that words are conventional signs of thoughts. This gives the reason for not choosing *sign* as the third term: a sign is not necessarily the expression of a concept but it is necessarily the sign for an object. For signs the *concept* vertex of the triangle is, as it were, optional whereas for expressions it is the *object* vertex which is optional. The following quotation from Husserl explains this in greater detail.

"Every sign is a sign for something, but not every sign has 'meaning', a 'sense' that the sign 'expresses'. In many cases it is not even true that a sign 'stands for' that of which we may say it is a sign. And even where this can be said, one has to observe that 'standing for' will not count as the 'meaning' which characterizes the expression. For signs in the sense of indications (notes, marks etc.) *do not express* anything, unless they happen to fulfill a significant as well as an indicative function. If, as one unwillingly does, one limits oneself to expressions employed in living discourse, the notion of an indication seems to apply more widely than that of expression, but this does not mean that its content is the genus of which an expression is the species. To mean is *not a particular way of being a sign in the sense of indicating something.* It has a narrower application because meaning—in communicative speech—is always bound up with such an indicative relation, and this in its turn leads to a wider concept, since meaning is also capable of occurring without such a connection. *Expressions* function meaningfully even in *isolated mental life, where they no longer serve to indicate anything.* The two notions of sign do not therefore really stand in the relation of more extensive genus to narrower species."[17]

This motivates the choice of *expression* as the third term of the threefold correspondence, giving the triple *object*, *concept*, and *expression.*

§ 2. The acts of the mind

The division of the acts of the mind into apprehension, judgement, and reasoning, is classical. Some variations can be found: Arnauld, for example, adds the fourth act ordering or method.[18] Simple apprehension, or perception, is an act of the mind in which the intellect comes to know something, as, for example, when seeing something.[19] The detailed study of apprehension belongs to psychology but the existence of this act is of importance also to logic since it provides the mind with raw material about which to think. The judgement will

[16] Arnauld, *The art of thinking*, Pt. 1, Ch. 4. Cf. Augustine, *De Dialectica.*

[17] Husserl, *Log. Unt. II*, Pt. 1, Inv. 1, § 1 (trans. Findlay).

[18] Arnauld, *The art of thinking*, p. 29. Also manifest from the division of his book into four parts: conception, judgement, reasoning, and ordering.

[19] Simple apprehension is the scholastic term used, e.g., by Gredt, *Elem. Phil.* n. 6. Perception is a modern equivalent used, e.g., by Locke, *An Essay Concerning Humane Understanding*, Bk. 2, Ch. 9.

Logic	(A) Acts of the mind by which something is understood, two in number:	(1) *Simple apprehension*, which is treated of by the doctrine handed down from Aristotle in the *Categories*;	
		(2) *Judgement*, in which is truth or falsity, which is treated of by the doctrine handed down from Aristotle in the book *Perihermeneias*.	
	(B) Acts of the mind by which reasoning proceeds from one to another, as regulated by *logic*. This act has three moods of procedure by which it deduces the conclusion, either necessary, probable, or false.	(1) *Analytic* or *judgemental logic*, which proceeds by resolution, treats of the minds moods of procedure which *induce necessity*, and can be considered in two ways:	(a) either from the point of view of the *form* of the syllogism, as is done in the book *Analytica Priora*;
			(b) or from the point of view of the *matter* of the syllogism, as is done in the book *Analytica Posteriora*.
		(2) *Inventive Logic* treats of the minds moods of procedure which *induce probability*, and is divided into three, according to what it generates: faith, suspicion, or appreciation:	(a) in *faith* and *opinion*, the mind is totally inclined towards one of two contradictories, but allows with dread for the other: and to this pertains the *Topics*;
			(b) in *suspicion*, the mind is not totally inclined towards either contradictory: and to this pertains the *Rhetoric*;
			(c) in *appreciation*, the soul is inclined towards one of the two contradictories because of its beautiful representation: and to this pertains the *Poetics*.
		(3) The part of logic which is called *Sophistry* treats of the minds moods of procedure which *induce error*, and Aristotle treats of this in the book *Elenchorum*.	

Table 1. A classical division of logic, as it appears in the works of Aristotle, according to the acts of the mind.

be treated of in Chapter II, Section 4, but, already at this point, a judgement can be defined as an act in which the intellect recognizes some form of agreement or discrepancy between concepts. Reasoning, treated of after the judgement, is an act of the mind by which, from known premisses, the mind comes to know a conclusion.

Table 1 shows an example of how logic used to be divided according the acts of the mind.[20] This division of logic persisted to the 18th and 19th century, as can be seen from the works of Kant and Bolzano.[21] This Table

[20] Author's translation of a table from the preface of Aquinas, 'In Perih.', p. ix.

[21] This can be established by looking at the table of contents of Kant, *Kritik der reinen Vernunft*, and Bolzano, *Wissenschaftslehre*.

also reveals two important points. First, that apprehension, inventive logic, and sophistry, no longer are considered part of logic proper, so the scope of logic has diminished. Second, that analytic logic is divided into formal logic and material logic. Formal logic studies the forms of correct thinking and material logic the content, truth, and necessity of the logical moods. Another way of characterizing the difference between formal and material logic is that formal logic studies the relation between concept and expression while material logic studies the relation between concept and object. Certainly there are pedagogic advantages to this division but, in a scientific treatment of logic, it seems to be a bad idea to take form and content apart.

Material logic, also called major logic, epistemology, or criteriology, can also be taken in a broader sense encompassing the material side of all parts of logic, not only of demonstration; this broadening of the scope of material logic originates with John of St. Thomas.[22] A typical division of material logic is into *proemial logic*, about the nature and definition of logic, the predicables and moods of predication; *predicamental logic*, about the content of apprehension, the categories, and the nature of universals; and *demonstrative logic*, about the content of judgement, reasoning, and demonstration.[23]

Broadly speaking, the main stream of logic has gradually turned from the principally material logic of the scholastic period to the prevailing formal logic, through the influence of logicians such as Leibniz, Boole, and Frege, culminating in the formalistic crown jewel *Principia Mathematica* by Whitehead and Russell published in 1910. On the other hand, formal logic was not unknown to the ancients as the content, and even the title, of Bocheński's book *Ancient Formal Logic* shows, and material logic is not entirely out of fashion, even though, perhaps, the name is.[24]

My opinion is that one has to maintain a certain balance between the formal and the material. That is, a complete method of logic must account both for the formal side of logic—how concepts are expressed—and for the material side of logic—how things are conceptualized. Instead of formal and material, one could use the modern counterparts syntactic and semantic. Thus, I will speak of a formal-material or, with a similar meaning, of a syntactic-semantic method of logic.[25]

§ 3. The principle of compositionality

Recall that an expression has been defined as a meaningful combination of words. How does such a meaningful combination of words arise? The first case is a word which is meaningful by itself, such as *man*, *two*, or *tiger*; such words are called *categorems*. The second case is that the expression consists of several words in a meaningful combination, such as *two plus three*, *white*

---

[22]Poinsot, *Material Logic*, pp. xv–xviii (Poinsot is the family name of John of St. Thomas).

[23]This division is after Gredt, *Elem. Phil.* n. 5.

[24]Since the definition of logic is a controversial matter, I have avoided it completely. Cf. Husserl, *Log. Unt. I*, § 3; Mill, *A System of Logic*, § 1; and Gredt, *Elem. Phil.* n. 4.

[25]The syntactic-semantic method of logic is associated with Martin-Löf. It should be noted that the words formal and syntactic also have modern senses, originating with Hilbert and Carnap respectively, according to which only that is formal or syntactic which treats of words without regard to meaning or content.

man, or *paper tiger*. Among these examples, the first two seem to be of a special nature since they are instances of the generally meaningful patterns *number plus number* and *white thing*. No such generally meaningful pattern exists in the third case. The pattern *something tiger* has meaningful instances like *white tiger* and *big tiger* but *paper tiger* is not one of them. The pattern *paper thing* also has meaningful instances such as *paper towel* and *paper box* but, again, *paper tiger* is not one of them. The difference between the two kinds of expression lies in how they are given meaning. In the first case, the meaning is *compositional*, i.e., the total meaning of the expression is somehow composed from the meanings of its parts. This is the normal case, as stated by Katz and Fodor in their seminal paper on semantics:

"Since the set of sentences is infinite and each sentence is a different concatenation of morphemes, the fact that a speaker can understand any sentence must mean that the way he understands sentences which he has never previously encountered is compositional: on the basis of his knowledge of the grammatical properties and the meanings of the morphemes of the language, the rules which the speaker knows enable him to determine the meaning of a novel sentence in terms of the manner in which the parts of the sentence are composed to form the whole."[26]

However, as demonstrated by the idiom *paper tiger*, not every expression has compositional meaning. In fact, idioms and irony generally have a non-compositional meaning. To treat of idioms and irony, however, lies outside the scope of the kind of logic dealt with here, which is why only expressions with compositional meaning will be considered in this book. With reference to Table 1, this means that I principally treat of analytic logic.

To recapitulate the above: an expression which consists of a single meaningful word is called a categorem; in the case of compositional meaning, one has to be able to identify the word which is used to give meaning to the whole expression, which is then called a syncategorem. For example, in the expression

two plus three times five,

the words *plus* and *times* are syncategorems and the words *two*, *three*, and *five* are categorems. This structure becomes apparent if the expression is displayed in tree-form:

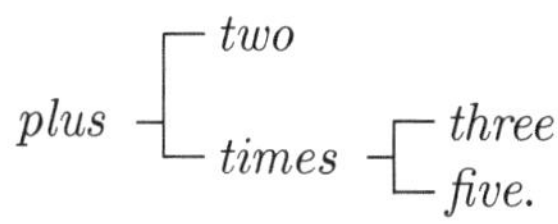

The categorems are the leaves of the tree and the syncategorems are the internal nodes. In computer science, this way of analysing an expression is called a parse tree or a syntax tree. Before a string of words becomes an expression, it is analysed into categorems and syncategorems. The definition of the notion of expression can now be made more exact by changing the genus from an

[26] Katz and Fodor, 'The Structure of a Semantic Theory', p. 171.

arbitrary string of words to a string of words analysed into categorems and syncategorems.[27]

The only thing demanded of categorems and syncategorems is that they be recognizable as instances of some abstract *form*. The categorems and syncategorems are words, printed on paper or spoken out loud, but their meanings are not in the concrete words but in the abstract forms to which we recognize that the words belong.[28] In the example above, *plus* is the form, with *two* and *three times five* as parts; continuing the analysis, *two* is a form without parts and *three times five* has *times* as form and *three* and *five* as parts. Every form has an arity which is zero or a positive integer: a form of arity zero corresponds to a categorem and a form of higher arity to a syncategorem. Forms are classified according to their arity into nullary forms, unary forms, binary forms, ternary forms, etc.[29]

Now we have the terminology in place to spell out the *principle of compositionality*: the meaning of a complex expression is determined by the meanings of its parts, together with a meaning contribution from the form. This principle is commonly attributed to Frege, even though it was not explicitly formulated by him.

Expressions can be either simple or complex and, according to the principle of compositionality, a complex expression has a complex meaning. As a slight digression, note that the principle of compositionality says nothing about the converse. That is, a categorem may well have complex meaning. This is the case, it seems, when we make abbreviatory definitions.

Even if the above analysis of an expression into form and parts is applicable to any expression with compositional meaning, the form of an expression may not always be apparent. In natural language, many different kinds of syntax are employed and, since this is an obstacle to analysis, one important purpose of a formal language is to make the forms of expressions apparent. Fortunately, the syntax of natural language is beyond the scope of this book.[30] It is useful to distinguish between grammatical form, i.e., the form that an expression has from the point of view of grammar, and logical form, i.e., the form discussed above, which is intimately connected to the meaning of the expression. With this distinction in place, we can write things, as we do, in natural language and understand them compositionally, without having to enter into lengthy discussions about how to parse natural language.

---

[27]Using the terminology of Cardelli and Wegner, 'On Understanding Types, Data Abstraction, and Polymorphism', § 1.1, our *untyped universe* consists of strings of words analysed into categorems and syncategorems, or, which amounts to the same, of syntax trees. A related notion from proof theory is that of a Herbrand universe (so called because Herbrand used it in his consistency proof presented in 'Sur la non-contradiction de l'arithmétique').

[28]In 'The theory of algorithms', nn. 5–7, p. 2, Markov makes the same distinction between what he calls *elementary signs* and the corresponding forms, which are called *abstract elementary signs*.

[29]When speaking *about* a form, one has to remember that an unsaturated form does not constitute an expression. When the arity has to be specified, the places where the parts go can be marked with dots, as in · *plus* · for example.

[30]For a connection between the syntax of natural language and type theory, vid. Ranta, *Type-Theoretical Grammar*.

§ 4. Lingua characteristica

Having established compositionality as the first principle of intuitionistic type theory, I will now look back on the historical development of related ideas. The pursuit of an exact universal language for science in general and mathematics in particular can be traced far back into the mists of time, perhaps all the way to Pythagoras. In fact, the Pythagorean view that concepts are in some way composed of numbers, or at least in the same way as numbers, is at the very backbone of this quest.[31] A prominent example of a philosopher attempting to exploit this Pythagorean idea, of which the principle of compositionality can be viewed as a modest form, was Lully, who invented a kind of mechanical device, called *Ars generalis ultima* or *Ars magna*,[32] by which subjects and predicates were fitted together to form philosophical and theological truths.[33] The ancient ideas were revived by the Neoplatonists in the 15th century: Ficino, Cusanus, da Vinci, and Copernicus, taking as their motto *omnia in numero et pondere et mensura disposuisti*, and continued in the 16th century by Galilei, and Kepler.[34] Another important contributor is Vieta with his analytic art which promises to leave no problem unsolved.[35] 17th century proponents of the program include Wilkins with his new symbolism,[36] Descartes, with whom the phrase *mathesis universalis* is strongly connected,[37] Weigel, who was Leibniz's mathematics professor, and, of course, Leibniz himself, who took up Lully's idea and called it *ars combinatoria*.[38] The next step for Leibniz was the *lingua characteristica universalis*, an ideographic language, where each concept is represented by one symbol. As Leibniz himself noted in his later years, the completion of the *lingua characteristica* will most probably remain a dream,[39] but the idea of such a language was later taken up by Frege in his *Begriffsschrift* and, more recently, by Leśniewski and Martin-Löf.[40]

Leibniz was not satisfied with a universal characteristic language, however, but thought that, once such a language was completed, one could find truth by merely computing with the symbols of the language. This calculus was called *calculus ratiocinator* and can be seen as the origin of symbolic logic. Leibniz's

---

[31] This view of the Pythagorean school is due to Aristotle, *Metaph.*, Bk. 13.

[32] These and the subsequent Latin phrases are translated as follows into English: *Ars generalis ultima*: The most general art, *Ars magna*: The great art, *mathesis universalis*: universal mathematics, *ars combinatoria*: the art of combination, *omnia in numero et pondere et mensura disposuisti*: thou hast ordered all things in measure, and number, and weight (Wis. 11: 21), *lingua characteristica universalis*: universal characteristic language, *Calculemus!*: Let us calculate!, *calculus ratiocinator*: calculus of reason.

[33] Turner, 'Raymond Lully'.

[34] Cf. Marciszewski, 'The principle of comprehension as a present-day contribution to mathesis universalis'; and Husserl, *The Crisis of European Sciences and Transcendental Phenomenology*, § 9.

[35] Vieta, *Artem Analyticam Isagoge seu Algebra Nova*. This book ends with the words *nullum non problema solvere*.

[36] Wilkins, *An Essay Toward a Real Character and a Philosophical Language*.

[37] 'Rules for the Direction of the Mind', Rule 4. Cf. the view on mathematics presented in *Discours de la Méthode*.

[38] Leibniz, *Dissertatio de Arte Combinatoria*.

[39] Letter to N. Remond, 1714, Leibniz, *Philosophical Papers and Letters*, p. 656.

[40] Leśniewski, 'Grundzüge eines neuen Systems der Grundlagen der Mathematik'; and Martin-Löf, *Intuitionistic Type Theory*.

optimism concerning *calculus ratiocinator* was most likely influenced by his interest in mechanical computers, first constructed by Pascal some 30 years before Leibniz presented his own design of such a device.[41] Thus *Calculemus!* was to be the answer to any dispute,[42] an idea which was not abandoned until it received the killing blow in the form of Gödel's incompleteness theorems.[43]

Symbolic logic was developed as a separate line of thought by contributors such as Jac. Bernoulli, Lambert, Ploucquet, De Morgan, Boole, Peirce, and Venn.[44] Other milestones in the development of the *lingua characteristica*, as a project distinct from symbolic logic, are Bolzano's investigations of the infinite, Peano's development of the axiomatic method, Hilbert's second problem, Russell's type theory, Brouwer's intuitionism, and Gentzen's natural deduction.[45] These topics will be developed and put into perspective in the course of this book.

The advent of digital computers has provided a strong incentive to the completion of the *lingua characteristica* because, as Perlis puts it, "within a computer, natural language is unnatural".[46] In fact, certain programming languages can be seen as modest realizations of it. A lot of present day work in computer science has some connection with the *lingua characteristica* and it is a curious fact that, in the past 50 years, computer science, as a subject, has had more impact on the *lingua characteristica* than logic, mathematics, and philosophy together. The reason seems to be that a certain amount of realism is, so to say, forced upon the computer scientist and this realism is, as it seems, an important ingredient in the *lingua characteristica*. This is why few implementations of mathematics on computer are based on Zermelo-Fraenkel set theory, despite it presently being the most popular foundation of mathematics.

Two programming languages, Iverson's APL and Martin-Löf's type theory, have been important sources of inspiration for this book.[47] Outstanding features of these languages are the clarity of meaning of type theory and the compactness of notation of APL. But this is the subject of later chapters of this book.

---

[41]Ifrah, *Räknekonstens kulturhistoria*, Vol. 2, pp. 444–446 (original in French: *Histoire universelle des chiffres*).

[42]Cf. Leibniz, 'Projet d'un art d'inventer', p. 176, cf. also Aristotle, *Pol.*, Bk. 1, Ch. 4, for a likely source of this dream.

[43]Gödel, 'Über formal unentscheidbare Sätze der Principia Mathematica und verwandter Systeme I'.

[44]A comprehensive list of references is given by Church, 'A bibliography of symbolic logic'. For the inevitable connection with computer science, cf. Dijkstra, 'Under the spell of Leibniz's dream'.

[45]Bolzano, *Paradoxes of the Infinite*; Peano, e.g., *Arithmetices Principia Nova Methodo Exposita*; Hilbert, e.g., 'Mathematical problems'; Russell, e.g., 'Mathematical Logic as Based on the Theory of Types'; Brouwer, e.g., 'Intuitionism and formalism'; and Gentzen, 'Untersuchungen über das logische Schließen I & II'.

[46]Perlis, 'Epigrams on Programming', n. 114.

[47]E.g., Iverson, *A Programming Language*; and Martin-Löf, 'Constructive mathematics and computer programming'. Perlis' remark that "a language that doesn't affect the way you think about programming, is not worth knowing" ('Epigrams on Programming', n. 19) applies to both of these languages. Other sources of inspiration have been the so called proof assistants: Automath, LCF, Mizar, HOL, PVS, NuPRL, Minlog, Coq, and Agda.

Leaving programming languages aside, this glance at the history of exact and formal languages shows the fine pedigree of our subject. It is both inspiring and discouraging that some of the greatest minds in the history of logic have attacked the problems confronting us with only a moderate amount of success.

CHAPTER II

Truth and Knowledge

THE THREEFOLD correspondence between things, thoughts, and words, discussed in the previous chapter, will now be investigated in further detail, with particular emphasis on mathematical entities; this investigation constitutes the first section of this chapter. In the next two sections, I attempt to show that common sense realism is not in conflict with intuitionistic type theory even though the latter *prima facie* seems to be a conceptualist framework. The judgement, and its syntactic counterpart assertion, are investigated in the fourth section. The fifth section treats of reasoning and the sixth section introduces the intuitionistic notion of proposition. In the seventh section, the laws of propositional logic are justified under the intuitionistic notion of proposition. The eighth section treats of schematic letters and variables. The ninth and last section treats of definitions.

The topics dealt with here are studied both in ancient (and medieval) philosophy, and in modern philosophy; but there is a certain tension between the two approaches. For example, when the ancients spoke about objects and propositions, they had in mind *men*, *horses*, and *this man is on the horse*. When modern philosophy speaks about objects and propositions, it has in mind *numbers*, *primes*, and *this number is prime*. In ancient and medieval philosophy the focus is on real things and the treatment of mathematical entities is often a kind of appendix, whereas in modern philosophy it is often the other way around. Since the *lingua characteristica* is supposed to be able to express propositions both concerning the real and the ideal, this tension has to be relieved. Consequently, the rather bold goal of this chapter is to provide a philosophical foundation for the *lingua characteristica* in general, and for intuitionistic type theory in particular, which accounts both for the real and for the ideal.[1]

§ 1. The meaning of meaning

Let it be laid down that the meaning of an expression is the concept expressed by it and that the referent of a concept, or of its expression, is the object signified. It is primarily the concepts which refer to their objects; the expressions refer only in a secondary sense: "an expression only gains an objective reference *because* it means something, it can rightly be said to signify or name the object *through* its meaning."[2] Sometimes the word denotation

[1]Further background can be found in Cocchiarella, 'Conceptual Realism as a Formal Ontology'.

[2]Husserl, *Log. Unt. II*, Pt. 1, Inv. 1, § 13 (author's translation). Cf. the parallel place in Aquinas, 'Summa Theol.', Pt. 1, q. 13, a. 1: "voces referuntur ad res significandas, mediante

J.G. Granström, *Treatise on Intuitionistic Type Theory*, Logic, Epistemology, and the Unity of Science 22, DOI 10.1007/978-94-007-1736-7_2,

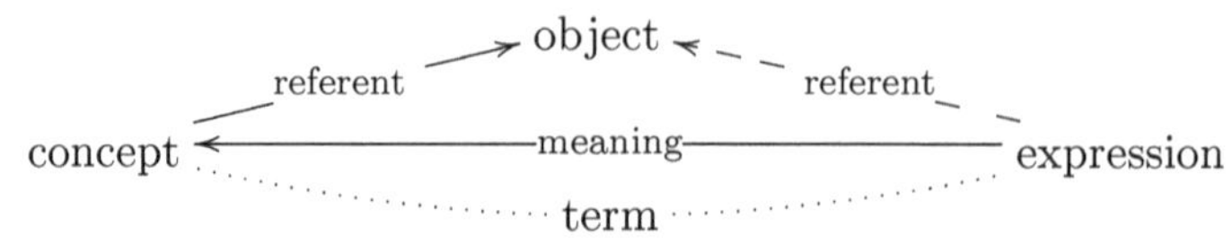

Figure 2. Meaning, referent, and term added to the triangle picturing the threefold correspondence between object, concept, and expression.

is used as synonymous with reference and an expression is said to *stand for*, signify, or name its referent. In the case of universal concepts, which refer to many objects, the objects are said to *fall under* the concept. One speaks about *the* referent of a concept with a unique object falling under it, and about *a* referent of a universal concept.

In addition to expression, meaning, and concept, logic also uses the word term.[3] Is the word term to be identified with expression, concept, or object? The classical definition of a term is that into which a predication can be analysed, namely, the predicate and the subject.[4] In the classical literature, the word term is used ambiguously between the expression and its meaning and, when a clarification is called for, the classical authors write *terminus scriptus* for the expression and *terminus mentalis* for the concept. In my opinion, the best way to understand the word term is as *an expression taken together with its meaning.* Cf. Figure 2. That is, it is neither the expression nor the concept, but both expression and concept taken together with the relation between them. One consequence of this is that for two terms to be equal, they have to have the same unambiguous expression. For example, even if the words *freedom* and *liberty* have the same meaning, they are considered distinct as terms. Moreover, ambiguous or equivocal words are considered different as terms when used in different senses in the same sentence.[5]

Terms sometimes refer to their objects *through another concept.* Compare for example *Paris* and *the capital of France.* The *meanings* of these two expressions are certainly very different. Let us agree to call a concept *immediate* if it signifies its object without any intermediate concept, such as *Paris*, and *mediate* if it signifies its object through some intermediary, as is the case with *the capital of France.*[6]

In intuitionistic type theory, a similar distinction is made between *canonical* and *noncanonical* terms or expressions.[7] In this setting, canonical corresponds

conceptione intellectus" : words refer to the thing signified, through the intellect's concept (author's translation).

[3]The Greek word ὅρος became *terminus* in Latin.

[4]Aristotle, *An. Pr.*, Bk. 1, Ch. 1, 24b17. Cf. Boëthius, 'De syllogismo categorico'.

[5]This use of the word *term* can be motivated as follows: the ancients speak about the three *terms* of a syllogism; *equivocation* is the fallacy of using an equivocal middle term in a syllogism, as in the argument "no light is dark; all feathers are light; therefore, no feathers are dark" (by Celarent); "an utterance is not called equivocal because it signifies many external things but because in signifying those many external things, there correspond to it different concepts in the soul." (Buridan, *Summulae de Dialectica*, Treatise 3, Ch. 1, § 2). Cf. the notion of polymorphism discussed on p. 92.

[6]Cf. Gredt, *Elem. Phil.* n. 16a.

[7]Cf. Martin-Löf, *Intuitionistic Type Theory*, p. 7.

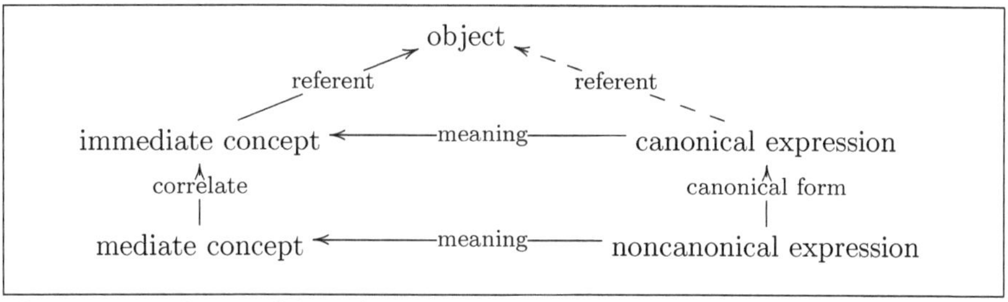

Figure 3. Mediate and immediate concepts compared with canonical and noncanonical expressions.

to immediate and noncanonical to mediate, though the names canonical and noncanonical apply to terms and expressions while mediate and immediate apply to concepts. The canonical term which corresponds to a noncanonical term is called its *value*. There seems to be no established terminology for the relation between an immediate concept and the corresponding mediate concept, so I think I will call the immediate counterpart of a mediate concept its *correlate*. The whole picture is given in Figure 3. Of course, the referent of a mediate concept is the same as the referent of its correlate, and the referent of a noncanonical expression is the same as the referent of its canonical form, which is the same as the referent of its meaning, i.e., of its concept.[8]

A correct understanding of the notion of *concept* is necessary for the correct understanding of the notion of judgement. The concept has variously been called *idea* or *species*,[9] *word of the mind* or *mental term*,[10] and *meaning* or *significatum*. The first pair of terms are derived from the verb *to see*; in this sense, the concept is the mental product of sight. Similarly, the word *concept* is itself derived from the verb *to conceive*. The second pair of terms draw on the analogy between concepts and words, while the third pair of terms are derived from the verbs *to mean* and *to signify* respectively, the former of Germanic and the latter of Latin origin. The doctrine of the concept as a *formal sign* captures all this.

"A formal sign is a sign whose whole essence is to signify. It is not an object which, having, first, its proper value for us as an object, is found, besides, to signify another object. Rather it is anything that makes known, before being itself a known object. More exactly, let us say it is something that, before being known as object by a reflexive act, is known only by the very knowledge that brings the mind to the object through its mediation."[11]

"We signify our concepts to others by spoken words. And that is so because in order to make known to others the very objects we know, we communicate to them the same means, the same formal sign, that we ourselves use to know

[8]To use the jargon of mathematical category theory, the diagram presented in Fig. 3 is *commutative*, i.e., following any chain of arrows from one point to another gives the same result.

[9]These two terms are translations of their Greek counterparts ἰδέα and εἶδος.

[10]The Latin expressions are *verbum mentis* and *terminus mentalis*.

[11]Maritain, *The Degrees of Knowledge*, Ch. 3, § 24.

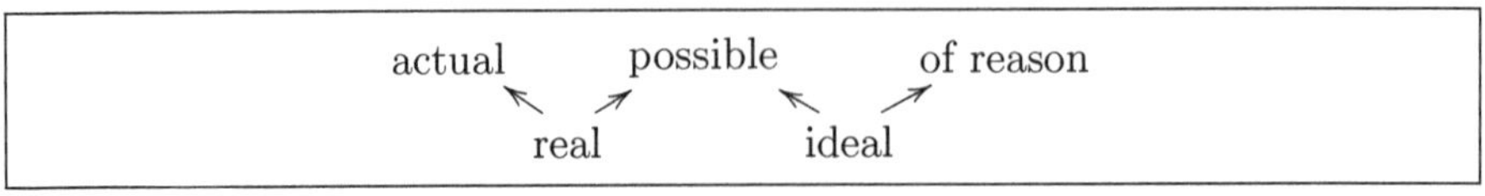

Figure 4. Division of modes of being: actual, possible, and of reason, into real and ideal.

these objects."[12]

It is with knowledge of concepts as with knowledge of grammar: they can be known on two levels. The concept can be known "by the very knowledge that brings the mind to the object through its mediation" just as grammar can be known as proficiency in the art of grammar. On the second level, the concept can be known "by a reflexive act", in the same way as grammar can be known through explicit knowledge of its laws, i.e., as a science. In the former case we speak about a *direct concept* and in the latter case we speak about a *reflex concept.*[13]

In general, when dealing with artificial languages, one makes a distinction between the language studied, the *object language*, and the language used to study the artificial language, the *metalanguage*. *Lingua characteristica*, or intuitionistic type theory, is our object language and ordinary English is our metalanguage. Typically, direct concepts are used in the object language and reflex concepts in the metalanguage. In Chapter III, reflex concepts will be brought into the object language.

§ 2. A division of being

Concepts normally refer to things and the first division of concepts is according to the things they refer to. Let us therefore turn our attention to things for a moment. Some things actually exist, such as animals and trees.[14] Other things have only possible existence, such as *a building five feet taller than the highest building in the world.* Moreover, things which have been actual, such as *a mammoth* or *Socrates*, are also called possible. Such things, possible or actual, are collectively called real beings—either because they are actual, because they can become actual, or because the have been actual.[15]

Yet another kind of being is that which is called a being of reason. Beings of reason cannot correspond to any thing, i.e., they cannot have any object *a parte rei*,[16] but exist only in the mind: "we say that these exist in the mind because the mind busies itself with them as kinds of being while it affirms or

---

[12]Maritain, *The Degrees of Knowledge*, App. 1, § 4, p. 419. That is, communication does not only consist in an exchange of words, but also of their meanings.

[13]Poinsot, *Material Logic*, p. 421. Cf. Husserl, *Log. Unt. II*, Pt. 1, Inv. 1, § 34; and Gredt, *Elem. Phil.* n. 16c.

[14]Cf. Aristotle, *Metaph.*, Bk. 7, Ch. 1, for the various senses of the word *being.*

[15]Strictly speaking, things which are actual are also called possible (*ab actu ad posse valet illatio*) so *real being* and *possible being* amount to the same; but, when real being is divided into actual and possible, possible has to be taken to exclude actual.

[16]*A parte rei*: on the side of things.

denies something about them".[17] Merely possible beings and beings of reason are collectively called ideal beings. Thus, an ideal being *does not* exist, whereas a being of reason *cannot* exist. The complete picture is given in Figure 4.[18] Note that a possible being is called both real and ideal.[19]

For example, blindness is a being of reason. To be blind means to not have sight. The concept blindness is formed from the concept sight by adding negation. Similarly with death, deafness, and other privations. Another kind of beings of reason are those which are a result of a formal abstraction, such as the line or circle of geometry or the numbers of arithmetic, which are totally devoid of sensible matter and thus cannot exist as things. This kind of being of reason is investigated below. Other examples of formal abstraction are the formation of the concept *redness* from *red*, *humanity* from *man*, etc. Yet another kind of being of reason are those of grammar and logic, such as *subject*, *predicate*, *proposition*, *set*, and *element*. Beings of reason are purely meaningful, or intelligible, entities, for which *the definition is everything*. How does the doctor confirm blindness in a patient? He checks for sight and when he does not find it he concludes blindness.

Of course, different beings of reason can be more or less distant from what is real. For example, a particular blindness is more real, more tangible, than, say, a particular prime number. In this sense, beings of reason admit of degrees in their distance from the real.[20]

According to Aristotle and his followers, the most basic concepts come from the direct apprehension of real being to the point that the nature, or species, of the thing is identified with the concept. This is the origin of the scholastic term *species expressa* for the concept. This nature, which in a sense is identified with the concept, is explained as follows by St. Thomas:

"Therefore, if it is asked whether this nature considered in this way can be said to be one or many, neither alternative should be accepted, because both are outside of the understanding of humanity, and either can pertain to it. For if plurality were included in its understanding, then it could never be one, although it is one insofar as it is in Socrates. Likewise, if unity were included in its notion and understanding, then Socrates and Plato would have one and the same nature, and it could not be multiplied in several things."[21]

Perhaps the analogy of a work of art will make this view clearer. Consider, e.g., Homer's *Iliad*. If you buy a copy of the *Iliad* in a bookstore, you get a piece of matter, paper, along with it and necessarily so, since the work itself

---

[17] Aquinas, *In Metaph.* Bk. 4, Les. 1, n. 12: "quam dicimus in ratione esse, quia ratio de eis negociatur quasi de quibusdam entibus, dum de eis affirmat vel negat aliquid" (trans. Rowan).

[18] After Maritain, *The Degrees of Knowledge*, Ch. 2, fn. 43.

[19] Cf. Husserl, *Log. Unt. II*, Pt. 1, Inv. 1, § 32: "Ideality in the ordinary, normative sense does not exclude reality" (trans. Findlay).

[20] Cf. Maritain, *The Degrees of Knowledge*, Ch. 2, § 33, p. 144.

[21] Aquinas, 'De ente et essentia', Ch. 2: "Unde si quaeratur utrum ista natura sic considerata possit dici una vel plures, neutrum concedendum est, quia utrumque est extra intellectum humanitatis et utrumque potest sibi accidere. Si enim pluralitas esset de intellectu eius, nunquam posset esse una, cum tamen una sit secundum quod est in Socrate. Similiter si unitas esset de ratione eius, tunc esset una et eadem Socratis et Platonis nec posset in pluribus plurificari." (Trans. Klima).

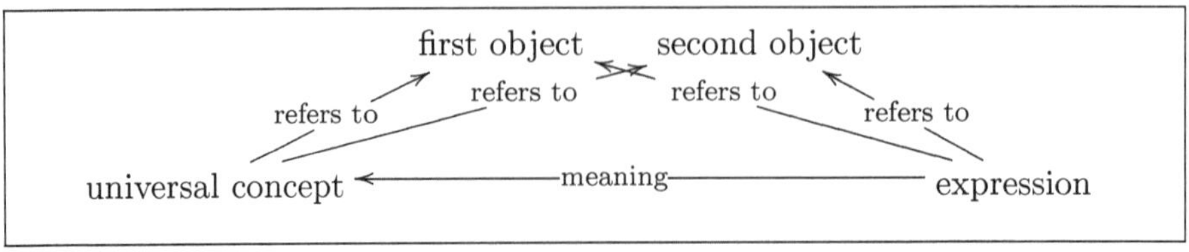

Figure 5. The threefold correspondence for universal concepts, which refer to many objects. In this figure there are two objects, but there can be arbitrarily many.

cannot be communicated but by the help of matter. Thus, the *Iliad* exists in your copy in the same way as the nature of a tree exists in the tree. Moreover, in one sense, it is the same work in different copies, and, in an analogous sense, it is the same nature in different trees.

§ 3. Mathematical entities

From the point of view of intuitionistic type theory, the most important entities are the mathematical entities. To make the transition from the real to the ideal less abrupt, let me take the following quotation from the first chapter of Biancani's *Treatise on the Nature of Mathematics* as a starting point.

"First we are going to discuss pure mathematics, i.e., geometry and arithmetic, which differs in kind from applied mathematics, namely, astronomy, optics, mechanics, and music. Quantity abstracted from sensible matter is usually considered in two ways. For it is considered by the natural scientist and the metaphysician in itself, that is, absolutely, insofar it is quantity, whether it is delimited or not; and in this way its properties are divisibility, locatability, figurability, etc. But the geometer and the arithmetician consider quantity not absolutely, but insofar as it is delimited, as are the finite straight or curved lines in continuous quantity; and the delimited surfaces from which there result various figures, like circle, triangle, etc.; and, finally, the solids, again delimited, which constitute the various species of solid figures, like pyramid, cube, cone, cylinder, etc., which pertain to the geometer. And the same can be observed analogically also in discrete quantity, i.e., in numbers, which the arithmetician considers only insofar as they are delimited. However, that it is these genera of delimited quantity that form the subject matter of geometry and arithmetic is clear from the fact that they define only these quantities, and they demonstrate only their various properties, which are entirely different from those that the natural scientist and the metaphysician consider in quantity absolutely. So it is obvious that these properties which the mathematician considers emanate from this quantity insofar it is delimited, such as equality, inequality, such and such division, transfiguration, various proportions, commensurability, incommensurability, construction of figures, etc. Obviously, these properties do not flow from the intrinsic nature of quantity, for if it is taken to be undelimited, the aforementioned properties do not follow, as nothing, taken to be like this, is equal or unequal, etc., but when delimitation is added to quantity, they flow from it by emanation. So it is correct to say that the formal aspect of mathematical consideration is delimitation, and that its total adequate object is delimited quantity, insofar as it is delimited. For from this delimitation there result various figures and numbers which the mathematician defines and of which he demonstrates various theorems. But this is the quantity that is usually called intelligible matter, in contradistinction to sensible matter, which concerns the natural scientist, for the former is separated by the intellect from the latter and it is perceived by the intellect alone. So continuous and discrete quantity, both delimited, are intelligible matter, the one of geometry, while the other of arithmetic. And from this it is also clear why the mathematician is said to consider finite quantity, for he considers delimited quantity, which is finite."[22]

This quotation expounds the view, formulated above, that mathematical entities are disengaged from sensible matter by a formal abstraction.[23] Any kind of mathematical modelling involves this kind of formal abstraction, both in natural science and in social science. To provide a simple example, consider a set of primary colours, e.g., red, yellow, and blue. If we introduce the abbreviations R, Y, and B for these colours we have done nothing mathematical. But if we forget what these letters stand for and treat them purely formally, then we have made a formal abstraction—we have disengaged from the real colours which these letters originally referred to. If instead R, Y, and B are abbreviations for names of persons, and we make a formal abstraction, the formalization will look the same.

When we make a formal abstraction, the entities considered gain a certain perfection—we idealize. The above quotation from Biancani continues as follows:

"Furthermore, as a result of mathematical abstraction from sensible matter, this abstract matter acquires a certain perfection, which is called mathematical perfection. For example, an abstract triangle is an absolutely plane figure constituted by three perfectly straight lines, by three angles, and by three absolutely indivisible point which, I think, could hardly be found in the nature of things (excepting, perhaps, celestial things). Hence many object to mathematicians that mathematical entities do not exist, except only in the intellect. However, we should know that even if these mathematical entities do not exist in that perfection, this is merely accidental, for it is well known that both nature and art intend to imitate primarily those mathematical figures, although because of the grossness and imperfection of sensible matter, which is incapable of receiving perfect figures, they do not achieve their end."[24]

This quotation shows that we can also consider mathematical entities without completely disengaging them from the real. In this sense, the perfect ideal line is said to be *possible in principle*, but because of "the grossness and imperfection of sensible matter" it does not exist in it.[25] I think this is how geometers

---

[22]Blancanus, 'A Treatise on the Nature of Mathematics along with a Chronology of Outstanding Mathematicians', p. 179. (Translation slightly modified by the author.)

[23]This abstraction is also labelled *pre-scientific* since it does not belong to mathematics proper, but is presupposed by it (Maritain, *The Degrees of Knowledge*, Ch. 2, § 22, cf. ibid., Ch. 4, § 6, p. 152).

[24]Blancanus, 'A Treatise on the Nature of Mathematics along with a Chronology of Outstanding Mathematicians', p. 180.

[25]Cf. Maritain, *The Degrees of Knowledge*, Ch. 4, §§ 6–7.

used to think about their subject matter before Gauss.[26] This view is no longer maintained in Euclidean geometry, but I think that it is tenable in arithmetic and that it still has a pedagogic value in geometry.

These distinctions provide the answer to an old puzzle, namely, whether all concepts are ultimately founded on real concepts according to the Peripatetic axiom *nihil est in intellectu quod non fuerit prius in sensu.*[27] The answer depends on the sense given to *founded on.* In a formal abstraction we disengage from the real, and ideal mathematical entities are *founded on* the real in the sense that they are the result of a formal abstraction from it. However, these are also *not founded on* the real in the sense that they are disengaged from it, i.e., in the sense that the *mathematical definition of number* does not contain any reference to reality.

This ideal quality of the mathematical concept of number should not lead us to believe that the connection to reality is of no importance. Take for example Lagrange's four square theorem, that every natural number n can be written as a sum of four squares

$$n = a^2 + b^2 + c^2 + d^2.$$

Having demonstrated this theorem, we want to be sure that all gravel in the nearest gravel-pit can be divided into four piles, each of which can be laid in a square. It is so because number is a being of reason founded on real being. Thus, while the *lingua characteristica* strictly speaking deals only with beings of reason, these beings of reason have to be founded on real beings; if they are not so founded, the whole project is reduced to inanity or mere navel-contemplation.

As said above, beings of reason do not have any object *a parte rei.* If we speak of an object for them, it is a purely formal or mathematical object. Cf. Figure 6. In this precise sense, intuitionistic type theory can be labelled a conceptualist framework. But, as I hope is clear from the above, there is no conflict between conceptualism for beings of reason and common sense realism. Thus, with respect to the age-old controversy between realists and conceptualists,[28] the present approach to *lingua characteristica* and intuitionistic type theory should be acceptable to both parties, as realists commonly agree that beings of reason have no object *a parte rei.*

The rejection of Platonic objects with extra mental existence does not make beings of reason into something subjective. Two senses of the word *objective* can be distinguished. The first and primary sense of the word is *on the object side* of the triangle. The second and derived sense is *the opposite of subjective*; it is derived because real objects are not subjective. Perhaps the second sense of the word objective is better described by the word *transsubjective.*[29] Mathematics

---

[26]Gauss came to the conclusion that geometry is not on a par with arithmetic in exactness (Maddy, 'Mathematical existence', fn. 18). Cf. Aristotle, *Metaph.*, Bk. 1, Ch. 2, § 5: "arithmetic is more exact than geometry". Also, with Einstein, it became clear that real space is not likely to be Euclidean (if that is taken to mean that the parallel postulate is valid).

[27]Aquinas, 'De Veritate', q. 2, a. 3, arg. 19. Author's translation: nothing is in the intellect that was not previously in the senses. Cf. Coffey, *The Science of Logic*, p. 7.

[28]Cf. e.g. Gredt, *Elem. Phil.* n. 114.

[29]Cf. Husserl, *The Crisis of European Sciences and Transcendental Phenomenology*, Pt. 2.

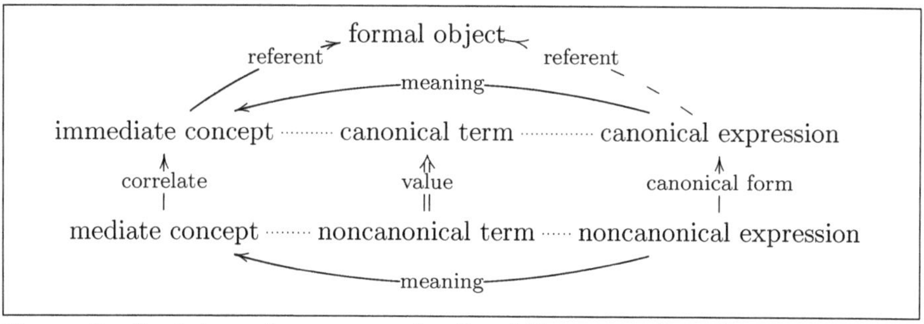

Figure 6. For beings of reason we get a threefold correspondence between expression, concept, and formal object.

is very objective in the second sense, but not so much in the first sense, since all its objects are purely formal.

In fact, all well-defined beings of reason are objective in the second sense because they are firmly founded in intelligible relations between concepts, or, to use Biancani's term, in intelligible matter. The whole of mathematics and the intuitionistic type theory developed in this book serve as examples of this objectivity. Husserl makes it clear that we can speak of meanings in themselves,[30] and the scholastic counterpart of these meanings in themselves is the *conceptus objectivus*, i.e., the concept taken in its objective, as opposed to mental, aspect.[31] In mathematics and other ideal sciences, we speak about formal objects; my opinion is that this talk about formal objects is to be understood as being tantamount to talk about objective concepts, the only difference lies in the vocabulary used to talk about them. When philosophers say that mathematics is founded in intelligible matter or intuition, they mean precisely that the formal object, the meaning in itself, or the objective concept, is objective in the second of the above two senses.[32]

This brings us to the important question of mathematical existence. That a certain expression is meaningful does not guarantee that its formal object *exists*. First, with Husserl, I like to make a distinction between *nonsense* and *absurdity*.[33] For example, we call *a largest prime number*, and *a square circle* absurd, but these expressions are still meaningful, i.e., they have a sense. If they did not have a sense we could not say that they do not exist.[34] Of course, the above manner of speaking about intelligible matter and intuition does not settle what makes an ideal object existent, as opposed to absurd.[35] There are three main answers to this question: Platonism, formalism, and intuitionism. The answer favored by intuitionistic type theory is, as hinted by its name, intuitionism; but let us investigate the three answers in order.

Platonism. The question of existence does itself not pose much of a problem

[30]Husserl, *Log. Unt. II*, Pt. 1, Inv. 1, § 35.
[31]Cf. Gredt, *Elem. Phil.* n. 7.
[32]Cf. Maritain, *The Degrees of Knowledge*, Ch. 4, § 6, pp. 152–154.
[33]Husserl, *Log. Unt. II*, Pt. 2, Inv. 6, § 12. Cf. ibid., Pt. 1, Inv. 1, § 15.
[34]Maritain calls absurd beings of reason the "thieves and forgers" among beings of reason (*The Degrees of Knowledge*, Ch. 2, § 33, p. 143).
[35]Cf. Bernays, 'Mathematische Existenz und Widerspruchsfreiheit'.

for Platonism as, according to this doctrine, the mathematical entities are as real as horses and elephants.[36] On the other hand, the distinction between *nonsense* and *absurdity* becomes problematic since, in its extreme form, Platonism is bound to claim that everything which is absurd is also nonsense: if sense entails existence, then lack of existence, i.e., absurdity, entails lack of sense. For example, it would be necessary to reject a geometrical figure that is both square and round as nonsense, since it does not exist.

Formalism. Formalism is associated with the idea that, if an object can be spoken about consistently, then it exists.[37] This view is motivated by certain mathematical insights of historical importance, e.g., that one can consistently add negative and irrational numbers to the language of arithmetic, because anything that can be demonstrated using them can also be demonstrated without using them. An objection against this view is that, for real being, consistency does not entail existence, so why should it do so for beings of reason? For example, one can consistently assume that there is intelligent life on another planet, since this assumption will never be refuted; but this does not entail that such life exists in the usual sense of the word.[38] On the other hand, in the sense that facts cannot contradict each other, existence entails consistency.

Intuitionism. Before treating of intuitionism, it is instructive to consider the point of view of finitism. The basic tenets of finitism are that all of mathematics should ultimately be founded on the natural numbers and that the natural numbers themselves are founded on the numerals.[39] Intuitionism is a refinement of finitism that adds an important ingredient, namely, the notion of *mental construction*.[40] The finitist view holds good only for canonical terms, but fails to explain noncanonical terms. In this book, mental constructions are identified with computations, treated of in Chapter IV. Intuitionistic type theory makes further refinements of intuitionism; these are explained further on in this book. The solution which I propose to the problem of existence will be given in connection with the investigation of definitions, in Section 9 of this chapter.

§ 4. Judgement and assertion

From grammar, we learn that a sentence is the verbal, oral or written, expression of a complete thought.[41] In writing, a sentence begins with a capital letter and ends with a period, question mark, or exclamation point. Having made this observation, the philosophically inclined is bound to attempt a classification of sentences. Accordingly, Diogenes reports that Protagoras divided discourse into four parts: entreaty, interrogation, answer, and injunction, and

---

[36]Cf. Maddy, 'Mathematical existence'.

[37]Hilbert, 'On the infinite', p. 370.

[38]Cf. Becker, 'Mathematische Existenz'.

[39]This view was expressed by Kronecker in the famous sentence "Die ganze Zahl schuf der liebe Gott, alles übrige ist Menschenwerk" (Cajori, *A History of Mathematics*, p. 362).

[40]This notion was introduced by Brouwer. Kant is the likely source of his terminology: "Philosophical knowledge is the knowledge gained by reason from concepts; mathematical knowledge is the knowledge gained by reason from the *construction* of concepts." *Kritik der reinen Vernunft*, Pt. 2.1.1, p. 469 (B 741) (trans. N. K. Smith).

[41]By sentence I mean that which was λόγος in Greek and became *oratio* in Latin.

that some writers claim that it was in fact seven parts, adding narration, promise, and invocation.[42] Aristotle mentions prayer as a kind of speech different from the enunciation.[43] Boëthius recognizes five kinds of sentences: questions, commands, invocations, deprecations, and enunciations.[44] St. Thomas has the same division as Boëthius, with only enunciations being true or false.[45] Buridan divides sentences according to their grammatical mood into indicative, imperative, optative, and subjunctive sentences.[46]

Modern philosophy is also interested in the classification of sentences under the heading of speech act theory.[47] Speech acts are often taken to include communication which is not verbal, such as the affirmative nod, so the notion of speech act is slightly more general than the notion of sentence pronunciation. Furthermore, in studying speech acts, the focus is shifted from the sentence to the act of pronouncing it. From this point of view, the spoken word is prior to the written word.[48]

According to the principle of compositionality, a sentence is analysed into form and parts. The outermost form of a sentence is called its logical mood, not to be confused with its grammatical mood. In his book *Expression and Meaning*, Searle makes a strong case for his fivefold division of the logical moods of sentences into:[49]

Assertives, which commit the speaker to some degree to some content. Searle gives the following examples: suggest, putting forward as a hypothesis, insist, flatly state, and solemnly swear.[50] Protagoras' narration and answer should fall under this heading as well as Boëthius' enunciations and Buridan's indicative sentences.

Directives, in which the speaker attempts to get the hearer to do something. This corresponds to Protagoras' interrogation, injunction, entreaty, and invocation; to Boëthius' questions, commands, invocations, and deprecations; and to Buridan's imperative and optative sentences. In addition to the above, Searle gives the following examples: order, request, beg, plead, pray, invite, permit, advise, dare, defy, and challenge.[51]

Commissives, which commit the speaker to some future action. Protagoras' promise belongs here, together with Searle's examples: vow, pledge, contract, guarantee, embrace, and swear.[52] A kind of commissive of particular importance later is the claim that a certain noncanonical term can be computed and the corresponding directive is the request for a term to be computed.

---

[42]Diogenes, *Lives of Eminent Philosophers*, Ch. 9.
[43]Aristotle, *Perih.*, Ch. 4.
[44]Boëthius, 'De syllogismo categorico', 767A.
[45]Aquinas, 'In Perih.', n. 85.
[46]Buridan, *Summulae de Dialectica*, Treatise 1, Ch. 2, § 3, p. 20.
[47]Cf. Austin's seminal book *How to Do Things with Words*.
[48]Cf. Aristotle, *Perih.*, Ch. 1.
[49]Searle, *Expression and Meaning*, pp. 12-20. Instead of logical moods of sentences, Searle speaks about kinds of "illocutionary acts".
[50]Ibid., p. 13.
[51]Ibid., p. 14.
[52]Ibid., p. 8.

Expressives, whose purpose is to express the speaker's psychological state about the content. Typical verbs include: thank, congratulate, apologize, condole, deplore, and welcome.[53] Although of great importance in ordinary life, this class has little or no relevance to the kind of logic we are interested in here.

Declarations, whose successful performance bring about the correspondence between the content of the declaration and reality. Some of Searle's examples are: *I resign*, *You're fired*, *I excommunicate you*, *I appoint you chairman*, *War is hereby dcclared*,[54] and when the judge says *You are guilty*.[55] Another kind of sentence which I think should be classified under this heading is the definition. Russell claims that definitions are not assertions, but volitions.[56] If this is correct, definitions should be classified under directives. On the other hand, it seems as the author of a text has the authority to define things as he see fit, within reasonable limits. Thus, I think that definition, as a legitimate exercise of authority, is to be classified under this heading.

The above four logical moods, excluding expressives, suffice for our logical and mathematical purposes. To logic, the most important mood is the assertive and many logicians consider no other kind of sentence. In this book, however, the other three forms of sentence are also considered.

Assertives form a rather large group of speech acts, and it has to be delimited somewhat to get the kind of sentence normally studied by logic. I will use the word *assertion* for the kind of speech act, or sentence, primarily studied by logic. The remark that it is studied by logic refers to the distinction between object language and metalanguage. We speak *about* assertions in the metalanguage and *make* assertions in the object language.[57] I take the force of an assertion to include assent, i.e., in making an assertion, the speaker subscribes to it; moreover, the assertive force of an assertion is always the normal indicative force.[58]

The word assertion is ambiguous between the act of assertion and its written representation, just as the word sentence is ambiguous between the speech act and the written representation of it. I will use the word judgement for the mental counterpart of an assertion, again ambiguously between the act of judging and the content judged.

We now come to a crucial point, namely the notions of *correctness* and *evidence* for judgements.[59] A judgement is first and foremost an *act* and, as act,

---

[53]Searle, *Expression and Meaning*, p. 15.

[54]Ibid., p. 16.

[55]Ibid., p. 19.

[56]Russell, 'The Theory of Implication', § 4.

[57]This remark is called for due to certain paradoxes which arise from subjecting assertive sentences uttered when speaking *about* logic to logical analysis. Cf. Russell, 'Mathematical Logic as Based on the Theory of Types', § 1.

[58]That I mention the *assent* as a separate component is with reference to the subtle but important distinction between neustic and tropic (Hare, 'Meaning and Speech Acts', § 4). I will, however, not make any further use of this distinction.

[59]I prefer the word correct to the word true to avoid confusion with true propositions, discussed later. Moreover, in connection with the extensional axiom of choice, Zermelo uses the same word, evidence (Ger. Evidenz), in a sense different from what I have in mind. In Zermelo's sense, that which is used extensively and, as it were, subconsciously

it has an agent; the content of a judgement is always evident *to* somebody.[60] When the content of a judgement is evident to somebody, it is nothing but a piece of his knowledge.[61] What a judgement means depends, of course, on the form of judgement. Thus, I have adopted Martin-Löf's rigid doctrine of *meaning explanations*; a meaning explanation always explains the meaning of a form of judgement, or assertion. To explain what a form of judgement means is the same as to explain under what circumstances one has the right to make the judgement.[62]

The content of the judgement is something objective, in the second of the above two senses, and the content is *correct* if it *can be made evident* to some intelligent being. Since evident and known are interchangeable in the sense explained above, this definition can be paraphrased by saying that the content of a judgement is correct if it is knowable.

This definition seems to indicate a reversal of priority between the objective and the subjective, because the objective correctness is defined in terms of the subjective evidence. Here I would like to make a distinction between judgements with *sensible* and *intelligible* matter: the former are dependent upon sense data while the latter are not. In the latter case this definition of correctness is not so controversial. Thus, consider an assertable content, e.g., *Socrates is a man* with sensible matter. Here Socrates, the substrate or suppositum, is objective in the first sense of the word; whereas *that Socrates is a man* is objective in the second sense of the word, i.e., transsubjective. My conclusion is that evidence is conceptually prior to correctness, whereas for judgements with sensible matter the substrate is temporally and ontologically prior to the judgement being evident.[63]

§ 5. Reasoning and demonstration

Reasoning is an act of the mind by which a certain judgement, the conclusion, is made evident. That is, the final act in a piece of reasoning is the act of judging its conclusion. The verbal expression of a piece of reasoning is called an argument, when dealing with reasoning in general, or a demonstration, when dealing with exact sciences.

A demonstration is analysed into inferences.[64] The mental counterpart of an inference brings the mind from certain judgements already known, the

---

by leading mathematicians is to be considered evident ('Neuer Beweis für die Möglichkeit einer Wohlordnung', p. 113).

[60]Cf. Martin-Löf, 'On the meanings of the logical constants and the justifications of the logical laws', p. 24.

[61]Ibid., p. 19.

[62]Evidence is not in things but in thought, is the type-theoretic counterpart of Aristotle's "truth and falsity are not in *things* but in *thought*" (*Metaph.*, Bk. 6, Ch. 4, § 2). Aristotle also makes clear that truth consists in the combination and separation of concepts in thought. Cf. Moore, 'The nature of judgement', p. 179.

[63]I take one thing to be ontologically prior to another if the latter cannot be conceived as existing without the former existing also.

[64]Martin-Löf, 'A Path from Logic to Metaphysics'; Sundholm, 'Inference versus Consequence'. Cf. also Aristotle, *An. Pr.*, Bk. 1, Ch. 1; *An. Post.*, Bk. 1, Ch. 10; *Top.*, Bk. 1, Ch. 1.

premisses, to a new judgement, the conclusion, which becomes known. I will write inferences in the form

$$\frac{P_1 \quad \cdots \quad P_n}{C} \; ,$$

where P_1 up to P_n are the premisses and C is the conclusion.[65] The premisses and the conclusion of an inference are always assertions.

More geometrico, a demonstration must start from premisses which are immediately known, without any need for further demonstration. Such an assertion is called an axiom, ἀξίωμα in Greek. In addition, certain inference steps are *immediate*, i.e., they do not admit further analysis. Instead of immediate, which is something negative, i.e., the absence of a means, one could say self-evident.[66] Thus, an assertion or inference is self-evident if it is "known by reason of the terms themselves, or by the explanation of the terms".[67] Instead of self-evident, it can be said to be evident *ex vi terminorum*, i.e., by force of the terms, or, which amounts to the same, *per se nota*, i.e., evident through itself. For immediate inferences, this means that, when the premisses are known, nothing more is called for to come to know the conclusion.[68] This explanation means that an axiom can be identified with an immediate inference with zero premisses.

This notion of self-evidence means that there may be some discourse which leads to the acceptance of a self-evident assertion or inference, viz., the explanation of the terms. This discourse is of course not demonstrative in the above sense of the word, though it may be termed apodictic in the derived sense of being necessary and absolute.[69] On the other hand, not every assertion accepted without discourse is self-evident. For example, assertions involving faith in a credible witness are accepted without discourse, but still not self-evident.[70]

§ 6. The proposition

In intuitionistic type theory, a distinction is made between an assertion and a proposition.[71] Although this distinction *prima facie* seems to be subtle and of little importance, it turns out to have far-reaching consequences. This

[65]Read an inference P_1 and $\cdots$ and P_n, *therefore* C. Instead of the horizontal line or the word *therefore*, one can use the sign ∴ (cf. Cajori, *A History of Mathematical Notations*, p. 667).

[66]Cf. Aristotle, *An. Post.*, Bk. 1, Ch. 2; Aquinas, 'In An. Post.', Bk. 1, Lect. 5; Poinsot, *Material Logic*, p. 461.

[67]Ibid., p. 462.

[68]This last explanation of what constitutes an immediate inference is due to Sundholm, 'Inference versus Consequence', p. 35. As an aside, in contrast to Whitehead and Russell, I do not think that an axiom can be accepted on purely practical grounds (cf. *Principia Mathematica*, Intro., Ch. 2, § 7, p. 62). The argument that "things have been taught to be self-evident and have yet turned out to be false" (ibid.) has little force, since, clearly, they were not self-evident after all: *errare humanum est*.

[69]Cf. Aristotle, *An. Pr.*, Ch. 1.

[70]Cf. Poinsot, *Material Logic*, p. 462.

[71]Subsequently I will prefer the word assertion to the word judgement. This choice differs from that of Martin-Löf, 'On the meanings of the logical constants and the justifications of the logical laws', who chooses *judgement* as the primary word, but it agrees with that of Russell, e.g., 'The Theory of Implication', § 1.1.

distinction is most clearly seen by an example where a proposition occurs unasserted. Let A and B be propositions, e.g., *the moon is a cheese* and *the moon is edible* respectively. Then *if A then B* is a new proposition, and in asserting that it is true, neither A, nor B, is asserted to be true. This is clear by the example. Geach calls this observation the Frege point, since Frege stressed it and made it explicit in the *Begriffsschrift.*[72]

Before defining what it means for something to be a proposition and what it means for a proposition to be true, two forms of assertion can be introduced, namely, that A is a proposition, which I will write

$$A : \text{prop},$$

and that a proposition A is true, which I will write

$$A \text{ true} .$$

That A is true presupposes that A is a proposition, since before we can know that a proposition is true, we must know that it is a proposition. The logical connectives operate on propositions. That is, granted that A and B are propositions,

$$A \,\&\, B, \quad A \vee B, \quad A \supset B, \text{ and } \Lambda$$

should also be propositions.[73] One of the first and most important tasks of a *lingua characteristica* is to explain the two forms of assertion, as well as the meanings of these connectives, and the two quantifiers, in such a way as to make the laws of propositional and predicate logic evident.

In the traditional approach to logic, the first division of propositions is into affirmations and denials. Here the word proposition is used in its traditional sense, corresponding to what I call an assertion. This symmetric treatment of affirmation and denial goes back to Aristotle and is founded on the law of excluded middle.[74] The modern version of this symmetry is the interpretation of a proposition, now in the modern sense, as a truth value, i.e., as referring to *the true* or *the false.*[75] Even if one agrees with the ontological version of the law of excluded middle, i.e., that propositions with real matter about the

---

[72]Geach, *Logic Matters*, p. 255. But, as pointed out by Klima, the Frege point was recognized long before Frege, for example, by Buridan, in *Summulae de Dialectica*, Treatise 5, Ch. 1, § 3, p. 308: "a syllogism has an additional feature in comparison to a conditional in that a syllogism posits the premises assertively, whereas a conditional does not assert them."

[73]These connectives are called conjunction, disjunction, implication, and *falsum* (or *absurdum*) respectively. Read them as follows in English: 'A and B', 'A or B', 'A implies B' (or 'if A then B'), 'not A', and '*falsum*' (or '*absurdum*'). The word connective applies strictly speaking only to the first three, since they connect A and B, but the meaning of the word is often extended to include *falsum* too (as well as negation and equivalence, see below). The symbol & is a ligature for the Latin word *et* meaning *and*; the symbol $\vee$ is just a stylized abbreviation of the Latin word *vel* meaning *or*; the symbol $\supset$ is due to Peano (*Arithmetices Principia Nova Methodo Exposita*, Log. Not., n. 2), in fact, $\subset$ is a stylized C abbreviating *is a consequence of*, so $B \subset A$ means that B is a consequence of A, or, equivalently, that A implies B; finally, the symbol Λ for *falsum* is due to Peano (ibid.), and it is a V for *verum* turned upside down.

[74]Aristotle, *Perih.*, Ch. 1 (cf. ibid., Ch. 4, 17a2).

[75]Boole, 'The Calculus of Logic'. Cf. Martin-Löf, 'On the meanings of the logical constants and the justifications of the logical laws', p. 14.

past or present are true or false, there are still problems with maintaining this symmetry between affirmation and denial.[76]

(a) The law of excluded middle is ontological, not logical. Bringing it into logic can be seen as an instance of the fallacy μετάβασις εἰς ἄλλο γένος.[77] I maintain that it is not a law of thought, i.e., a law of logic, but a principle of being.

(b) Although the law of excluded middle has a kind of intuitive validity for real being, it is not evident for beings of reason.[78] Should not the laws of logic hold for pure mathematics?

(c) Many predicates in natural language are vague and allow for borderline cases.[79] Such predicates do not fare well in classical logic but are treated of without problems in intuitionistic logic, where the law of excluded middle is not accepted as a law of thought.

(d) The laws for forming propositions by quantification over infinite domains are difficult to justify under the classical interpretation of a proposition as a truth value:[80] indeed, as pointed out by Brouwer in 1908, it is not even likely to be possible.[81]

So, what does intuitionism suggest instead of the definition of a proposition as a truth value? Put differently, what does the form of assertion A : prop mean?

Definition 1. *A proposition is defined by laying down what counts as a cause of the proposition.*[82]

With this definition in place, it is natural to define truth of a proposition in the following way.

Definition 2. *A proposition is true if it has a cause.*

To understand the word *cause* in these definitions, consider the classical dictum *scire est rem per causas cognoscere*.[83] This notion of truth of a proposition has Leibniz's principle of sufficient reason, as it were, built in. The

---

[76]The precise sense in which the law of excluded middle can be considered ontological is clarified in Ch. VI, § 2.

[77]I.e., the jumping into a different domain or science. The phrase is derived from Aristotle, *An. Post.*, Bk. 1, Ch. 7, 75a38, which is concerned with the impossibility of proving facts in one science using the methods of another, e.g., to prove a geometrical fact by appeal to optics.

[78]Husserl, *Log. Unt. II*, Pt. 2, Inv. 6, § 30.

[79]Cf. Geach, 'The law of excluded middle', pp. 71-73.

[80]Martin-Löf, *Intuitionistic Type Theory*, p. 11.

[81]Brouwer, 'The Unreliability of the Logical Principles'.

[82]This definition, and the following, is a copy of Martin-Löf's definition (*Intuitionistic Type Theory*, p. 11) with the word *proof* replaced by the word *cause*.

[83]To know is to have cognizance of the thing through causes. This dictum is derived from Aristotle, *An. Post.*, Bk. 1, Ch. 2, 71b9, sqq. Cf. *Metaph.*, Bk. 2, Ch. 1, n. 5, sqq. Other formulations are the poetic "Felix, qui potuit rerum cognoscere causas" (Virgil, *Georgics*, Bk. 2, l. 490) and "Vere scire, esse per causas scire" (Bacon, *Novum Organum*, Bk. 2, Ch. 20). With respect to the division of causes (Aristotle, *Metaph.*, Bk. 5, Ch. 2; *Phys.*, Bk. 2, Ch. 3), the kind of cause I have in mind here could be called a *logical cause* (cf. *An. Post.*, Bk. 2, Ch. 11). The relation between truth and causes will be further developed in connection with the law of excluded middle, treated of in Ch. VI, § 2.

a cause of	consists of
$A \& B$	a cause of A and a cause of B;
$A \vee B$	a cause of A or a cause of B, together with information about which cause it is that is given;
$A \supset B$	a method which takes any cause of A into a cause of B;
Λ	(there is no cause of Λ).

Table 2. The intuitionistic interpretation of the propositional connectives, i.e., the BHK interpretation.

principle of sufficient reason is that "in virtue of which we hold that, no fact can be found true, nor can truth exist in any proposition, unless there be a sufficient reason, why it is so rather than otherwise, although these reasons most often cannot be known by us."[84]

Thus, when I say that I know that the proposition A is true, I mean that I am in possession of a cause of it. In this setting, the cause could also be called a *reason*,[85] i.e., the reason by which I know that A is true. The distinction between cause (*causa*) and reason (*ratio*) is a virtual distinction: a cause is taken as an *objective* ground of a proposition whereas a reason is taken as a *particular subject's* ground for holding the proposition to be true.[86]

These definitions, of proposition and truth, are of dubious value until it becomes clear that all classical laws of logic, except the law of excluded middle, can be justified from them by assigning suitable meanings to the logical connectives. The intuitionistic interpretation of the propositional connectives is given in Table 2.[87] Since a proposition is defined by laying down what counts as a logical cause of it, and this is laid down in Table 2, the inference rules[88]

$$\frac{A : \text{prop} \quad B : \text{prop}}{A \,\&\, B : \text{prop}}\,, \qquad \frac{A : \text{prop} \quad B : \text{prop}}{A \vee B : \text{prop}}\,,$$

[84]Author's translation of Leibniz, 'Monadologia', § 32: "vi cujus consideramus, nullum factum reperiri posse verum, aut veram existere aliquam enunciationem, nisi adsit ratio sufficiens, cur potius ita sit quam aliter, quamvis rationes istæ sæpissime nobis incognitæ esse queant."

[85]It is difficult to determine to what extent Leibniz identified *ratio* with *causa*; cf. Di Bella, 'Causa Sive Ratio'.

[86]In this setting, it also makes sense to call the cause or reason a *truth-maker*, since, in a sense, it is the cause that makes the proposition true. Cf. Sundholm, 'Existence, Proof, and Truth-Making: A Perspective on the Intuitionistic Conception of Truth'.

[87]Martin-Löf, *Intuitionistic Type Theory*, p. 12. This interpretation is called the BHK interpretation after its discoverers Brouwer (in many of his works), Heyting ('Sur la logique intuitionniste'), and, independently, Kolmogorov ('Zur Deutung der intuitionistischen Logik'). It should be mentioned that there is direct line of thought from Husserl to the BHK interpretation: Becker, one of Husserl's students, interpreted propositions as expectations ('Mathematische Existenz'), and influenced Heyting who interpreted propositions as problems (cf. Mancosu, *From Brouwer to Hilbert*, pp. 275–285). This leads to the identification of: (1) the cause of a proposition, (2) the fulfillment of an expectation, and (3) the solution of a problem.

[88]Note that a distinction is made between an *inference* and a *rule of inference*. Exactly what it means for a rule of inference to be valid is explained in § 8 of this chapter.

and

$$\frac{A : \text{prop} \quad B : \text{prop}}{A \supset B : \text{prop}}$$

are self-evident, and so is the axiom

$$\Lambda : \text{prop}.$$

There are two connectives missing from this list, namely, negation and equivalence. These connectives can be defined in terms of the already introduced connectives by nominal definition. The negation of a proposition A is written $\sim A$ and defined by

$$\sim A \stackrel{\text{def}}{=} A \supset \Lambda : \text{prop}.$$[89]

This definition of negation is commonly accepted in intuitionistic logic,[90] though other definitions have been proposed in other areas of logic. Equivalence between two propositions A and B is written $A \supset\subset B$ and defined by

$$A \supset\subset B \stackrel{\text{def}}{=} (A \supset B) \,\&\, (B \supset A) : \text{prop}.$$[91]

This definition of equivalence seems to be universally accepted.

Having so defined the notion of assertion and explained the first two forms of assertion, namely, A : prop and A true, a distinction is to be made between a *complete* and an *incomplete* assertion.[92] The form of assertion

$$A \text{ true}$$

is incomplete in the sense that it suppresses the cause. I will write c : cause(A) if c is a cause of A.[93] The meaning of the form of assertion A : prop is that it has to be laid down what counts as a cause of it. That is, a proposition A is defined by defining the form of assertion c : cause(A). Since that a proposition is true means that it has a cause, the inference rule

$$\frac{c : \text{cause}(A)}{A \text{ true}}$$

is self-evident and completely determines the meaning of the form of assertion A true.

The forms of assertion c : cause(A) and A : prop are both complete. Indeed, they are examples of the first form of complete assertion, the predication, where

[89]The symbol $\sim$ for negation is due to Russell ('Mathematical Logic as Based on the Theory of Types', § 6).

[90]But cf. Bishop and Bridges, *Constructive Analysis*, pp. 10–11.

[91]The symbol $\supset\subset$ for equivalence is due to Heyting ('Die formalen Regeln der intuitionistischen Logik', § 2).

[92]An incomplete assertion, e.g., A true, constitutes an *incomplete communication* (unvollständige Mitteilung) in that the speaker suppresses certain information (cf. Hilbert and Bernays, *Grundlagen der Mathematik*, p. 33; and Kleene, 'On the Interpretation of Intuitionistic Number Theory', § 1). Also, what I call an incomplete assertion was called a *judgement abstract* (Urteilsabstrakt) by Weyl ('Über die neue Grundlagenkrise der Mathematik', p. 54).

[93]This important step of bringing the causes into the language of logic, i.e., of naming them, was first taken by Martin-Löf, 'An intuitionistic theory of types', p. 77, under the guise of proof objects. Cf. Martin-Löf, 'Analytic and synthetic judgements in type theory', where the distinction between the complete assertion c : cause(A) and the incomplete assertion A true is related to the Kantian distinction between analytic and synthetic judgements.

something, the predicate is predicated of something, the subject. In predication, the connection between the subject and the predicate is expressed by the copula, which is the present tense of the verb 'to be', the verb substantive.[94] In intuitionistic type theory, the copula is often spelled colon which is read *is*.

Examples of predicates are 'prop' and 'cause(A)', for a proposition A. To get another example, define a number, in the sense of Peano,[95] to be either zero or the successor of a number. If we write 0 for zero and $\mathrm{s}(a)$ for the successor of a, we get the axioms

$$0 : \text{number}$$

and

$$\frac{a : \text{number}}{\mathrm{s}(a) : \text{number}} . \tag{1}$$

This makes 'number' a third example of a predicate.[96] Of course, propositions can involve numbers in the usual way. If $a < b$ is defined by stipulating that $a < \mathrm{s}(a)$ for any number a, and that if $a < b$, then $a < \mathrm{s}(b)$, then the inference rule

$$\frac{a : \text{number} \quad b : \text{number}}{a < b : \text{prop}} \tag{2}$$

becomes evident since $a < b$ is defined as a proposition.[97] Moreover the inference rules

$$\frac{a : \text{number}}{a < \mathrm{s}(a) \text{ true}} \tag{3}$$

and

$$\frac{a < b \text{ true}}{a < \mathrm{s}(b) \text{ true}}$$

become evident in virtue of the definition.

I called the predication $a : P$ the first form of complete assertion. The second form of complete assertion is the assertion of definitional equality. It turns out to be a bad idea to treat of equality in the general form

$$a = b,$$

because we first have to spell out what kind of objects a and b are, and, in this general form of equality, there is no guarantee that a and b have a common genus.[98] On the other hand, if we already know that $a : P$ and $b : P$ for some predicate P, then this form of assertion has good sense, and can be written

$$a = b : P$$

---

[94]Many assertions can be put into this form, e.g., "a man walks" means the same as "a man is walking". Cf. Aristotle, *Perih.*, Ch. 12.

[95]Peano, *Arithmetices Principia Nova Methodo Exposita*, § 1, with the difference that, as is now customary, the first number is zero instead of one. I find it more natural to start the number series in the sense of Peano at zero since, if starting at one, there are two different formalizations of the unit, the starting point *one*, and the s for the successor.

[96]For the moment, I ignore the distinction between canonical and noncanonical terms.

[97]With mention of the causes, the definition of $a < b$ becomes: there is a cause of $a < \mathrm{s}(a)$, and if there is a cause of $a < b$, then there is a cause of $a < \mathrm{s}(b)$.

[98]Geach, 'Identity', p. 3. Cf. Quine's dictum: "no entity without identity" (*Theories and Things*, p. 102).

so as to explicitly show what kind of objects a and b are. That which stands on the right-hand side of the colon, i.e., the predicate P above, will be called a *logical category*.[99] In intuitionistic type theory, every logical category comes equipped with a definitional equality. I have not yet defined what definitional equality between propositions and causes mean—the complete definition will have to wait until Chapter IV—but, already at this point, it can be said that the relation of definitional equality between objects of any logical category should satisfy

(1) that the two terms of a definition are equal,

(2) that equals can be substituted for equals giving equal results,

(3) that any object is equal to itself, and

(4) that two objects which equal a third are equal to one another.[100]

An example of (3) is the assertion $0 = 0 : \text{number}$, and an example of (2) is the inference rule

$$\frac{a = b : \text{number}}{\mathrm{s}(a) = \mathrm{s}(b) : \text{number}} .$$

When defining things in the way we are used to in mathematics, we use definitional equality. For example, when addition between numbers is defined by the two equations

$$\begin{cases} a + 0 & = & a & : \text{number}, \\ a + \mathrm{s}(b) & = & \mathrm{s}(a + b) & : \text{number}, \end{cases}$$

the two sides of the equality sign are definitionally equal. To express this in inference rules, first note that the above definition of addition makes evident the inference rule

$$\frac{a : \text{number} \quad b : \text{number}}{a + b : \text{number}} ,$$

because $a + b$ can always be computed by the above equations. Moreover, the two inference rules

$$\frac{a : \text{number}}{a + 0 = a : \text{number}}$$

and

$$\frac{a : \text{number} \quad b : \text{number}}{a + \mathrm{s}(b) = \mathrm{s}(a + b) : \text{number}}$$

are evident from the definition of addition. Similarly, the definition

$$\begin{cases} a \times 0 & = & 0 & : \text{number}, \\ a \times \mathrm{s}(b) & = & (a \times b) + a & : \text{number}, \end{cases}$$

[99] This use of the word category is looking more towards Kant than to Aristotle (cf. Coffey, *The Science of Logic*, p. 151). In fact, Kant's categories are more or less his forms of judgement and this fits rather well with the above definition of a logical category. The word category was also used in the sense of logical category by de Bruijn (*Automath, a language for mathematics*, § 1.2.1).

[100] Cf. Martin-Löf, 'About models for intuitionistic type theories and the notion of definitional equality', p. 93; and Ch. III, § 3, p. 59, sqq., of this book.

of multiplication makes evident the inference rule

$$\frac{a : \text{number} \quad b : \text{number}}{a \times b : \text{number}},$$

and the two inference rules

$$\frac{a : \text{number}}{a \times 0 = 0 : \text{number}}$$

and

$$\frac{a : \text{number} \quad b : \text{number}}{a \times \text{s}(b) = (a \times b) + a : \text{number}}.$$

The definitional equality mentioned above is not in every way tantamount to the usual mathematical equality. In mathematics, some equalities are definitional and some are not. For example, when we prove some equality by mathematical induction, e.g., that addition is commutative,

$$a + b = b + a,$$

the equality is not definitional. This is because by proving it by induction we give the cause of the two terms being equal, i.e., this equality has to be expressed by a proposition. Consequently, a distinction has to be made between definitional equality and propositional equality: definitional equality is a complete form of assertion whereas propositional equality is a form of proposition. That two numbers a and b are propositionally equal will be written a eq b,[101] and this is a proposition, i.e.,

$$\frac{a : \text{number} \quad b : \text{number}}{a \text{ eq } b : \text{prop}}.$$

A cause of two numbers being propositionally equal is existent if they are definitionally equal, i.e.,

$$\frac{a = b : \text{number}}{a \text{ eq } b \text{ true}}.$$

That addition is commutative is now expressed by the incomplete assertion

$$(a + b) \text{ eq } (b + a) \text{ true},$$

[101] I use the standard equality sign = for definitional equality. This sign was introduced by Recorde, *The Whetstone of Witte*, in 1557: "And to avoide the tediouse repetition of these woordes: is equalle to: I will sette as I doe often in woorke use, a paire of paralleles, or Gemowe lines of one lengthe, thus: =, bicause noe 2 thynges, can be moare equalle." (there are no page numbers in this work, but the quoted passage stands under the heading "The rule of equation, commonly called Algebers Rule" which occurs about three quarters into the work). This use of the equality sign seems to me most natural since we use it when we make abbreviatory definitions in mathematics. Thus I had to use another sign for propositional equality. In the type-theoretic literature, there are several suggestions, including 'I' (Martin-Löf, *Intuitionistic Type Theory*, p. 59), and 'Id' vs. 'Eq' (with a slight difference in meaning, Nordström, Petersson and Smith, *Programming in Martin-Löf's Type Theory*, Ch. 8). According to Cajori ('Mathematical Signs of Equality', p. 116), the most popular notation, both before Recorde and in competition with him, was to write equality in words, i.e., something like "æquales", "égale", "gleich", or the abbreviation "æq".

i.e., the proposition $(a + b)$ eq $(b + a)$ is found to be true by finding a cause of it. Propositional equality can be negated, e.g., one of Peano's axioms for arithmetic is

$$\sim(0 \text{ eq } \mathrm{s}(0)) \text{ true} .$$

A proof of this axiom is given by Martin-Löf in *Intuitionistic Type Theory* (p. 91).[102] Since an assertion cannot be negated, this shows another difference between propositional and definitional equality.

A form of assertion typically has a number of *presuppositions* which are assertions which must be known in order for it to make sense.[103] In everyday language, presuppositions are most easily observed in sophistical questions, like *Do you still beat your wife ?*, which can neither be affirmed, nor denied, unless the presupposition is fulfilled. We have already seen that the forms of assertion A true and $c : \mathrm{cause}(A)$ presuppose that A is a proposition and that $a = b$: number presupposes that a and b are numbers. I will also use the word presupposition in a more general sense, according to which more specific forms of assertion can have more specific presuppositions. For example, that A & B is a proposition *presupposes*, in this more general sense, that A and B are propositions, because one cannot come to know that A & B is a proposition except by first knowing that A and B are propositions. Strictly speaking, inference rules where the conclusion is a presupposition of the premiss, in either sense, like

$$\frac{A \text{ true}}{A : \text{prop}}$$

and

$$\frac{A \mathbin{\&} B : \text{prop}}{B : \text{prop}} ,$$

are *valid* but *useless* ; the conclusion is already known before the premiss, so there is no point in inferring it.

An inference rule is called *well-formed*, if all presuppositions of the conclusion C can be inferred from the premisses P_1 up to P_n taken together with their presuppositions. When accepting a premiss, one also implicitly accepts its presuppositions, and if the presuppositions themselves have presuppositions, these are also accepted, etc. ; thus one may spell out the presuppositions recursively. The relation of well-formedness imposes an order on the inference rules, because the validity of other inference rules may be needed to show that a particular inference rule is well-formed. For example, the conclusion of inference rule (3) presupposes that $a < \mathrm{s}(a)$ is a proposition ; this is demonstrated from the premiss a : number by

$$\frac{a : \text{number} \quad \dfrac{a : \text{number}}{\mathrm{s}(a) : \text{number}}}{a < \mathrm{s}(a) : \text{prop}} ;$$

thus, inference rules (1) and (2) have to come before inference rule (3). Further examples of this phenomenon will be seen later.

[102] Thus, properly speaking, this is not an axiom, but a theorem, of intuitionistic type theory.

[103] The first detailed analysis of the notion of presupposition was given by Duns Scotus, 'De rerum principio'.

Recall that an assertion is the expression of a judgement and that a demonstration is the expression of a piece of reasoning. Furthermore, recall that a judgement is correct if it can be made evident, i.e., if it is knowable. With respect to the relation between demonstrability and correctness of an assertion, two questions can now be formulated.[104]

(1) Is every demonstrable assertion correct?

(2) Is every correct assertion demonstrable?

That the answer to (1) is vehemently *yes* follows from what demonstration and correctness mean. If you reason according to valid inference rules you arrive at knowledge of the conclusion, whence the conclusion is correct. That is, intuitionistic type theory is *sound* since its inference rules are made evident. The answer to (2) depends on whether we are confronted with a complete or an incomplete assertion.

In the second case, the answer is *no*, according to Gödel's incompleteness theorem, if we consider the inference rules to be fixed.[105] If we allow new inference rules to be justified and added to the logical system, the answer becomes *yes in principle*.[106] I say *in principle* because, as indicated by Leibniz's principle of sufficient reason (p. 29), in many cases, the causes cannot, practically speaking, be known to us.

For complete forms of assertion, the answer is again *yes in principle* if we allow new inference rules to be justified and added to the logical system. Thus, if *demonstrable* is taken to mean demonstrable by any valid inference rules, not fixed in advance, then *demonstrable* and *correct* coincide. A more interesting question is whether the inference rules of intuitionistic type theory (e.g., those presented in this book) form a complete and finished system of inference rules for its forms of expressions.

Thesis 1. *Any correct and complete assertion expressible in a certain part of intuitionistic type theory can be demonstrated by means of the inference rules justified in that part of intuitionistic type theory.*[107]

---

[104]These two questions are the type-theoretic equivalents of what Quine calls *soundness* and *completeness* for a system of logic ('A proof procedure for quantification theory', p. 145).

[105]Gödel, 'Über formal unentscheidbare Sätze der Principia Mathematica und verwandter Systeme I'. More specifically, fixing a collection of forms of expression and their corresponding inference rules, containing the expressions and rules of arithmetic, there are arithmetic propositions which cannot be demonstrated using only inference rules from this collection, but which are demonstrable, and hence correct, using valid inference rules outside of the collection.

[106]Cf. Martin-Löf, 'On the meanings of the logical constants and the justifications of the logical laws', p. 37. Cf. Ch. VI, § 2, of this book. Note that, unless we fix our inference rules, the answer to (2) cannot be *no*. To answer *no* we have to know an assertion to be correct but not demonstrable, but the only way to come to know that an assertion is correct is though a demonstration.

[107]Cf. Martin-Löf, 'Analytic and synthetic judgements in type theory', p. 97. Note also that it can mechanically be decided whether or not an assertion is demonstrable from the inference rules of a certain part of intuitionistic type theory (cf. e.g. Coquand, 'An algorithm for type-checking dependent types'), i.e., there is a kind of bivalence for assertions with respect to demonstrability.

Some evidence for this thesis is given by the detailed justifications of the inference rules in the subsequent presentation of intuitionistic type theory. By *a certain part of intuitionistic type theory* I mean intuitionistic type theory with a certain number of constructions of sets admitted. Note that the correctness of Thesis 1 is not crucial—it will never be referred to in the justification of something else.

§ 7. The laws of logic

The notions of proposition and truth were defined above, together with the logical connectives &, $\vee$, $\supset$, Λ, $\sim$, and $\supset\subset$. It is now time to fulfill the promise that all the standard laws of propositional logic, except the law of excluded middle, can be justified under this intuitionistic interpretation of the aforementioned notions. Here I have used the phrase law of logic as more or less synonymous with inference rule; this is clearly an oversimplification of how the phrase is usually understood, but it fits well with intuitionistic type theory.

The connectives will now be considered in order. First out are the inference rules governing conjunction. The inference rule

$$\frac{A \text{ true} \quad B \text{ true}}{A \,\&\, B \text{ true}}$$

is self-evident upon remembering that a cause of A & B consists of a cause of A and a cause of B. Note that this inference rule is *well-formed* since the presupposition of the conclusion, that A & B is a proposition, follows from the presuppositions of the premisses, i.e., that A and B are propositions. Similarly, the inference rules

$$\frac{A \,\&\, B \text{ true}}{A \text{ true}}$$

and

$$\frac{A \,\&\, B \text{ true}}{B \text{ true}}$$

are self-evident. They are well-formed because that A & B is a proposition presupposes that A and B are propositions. Moreover, the truth of a conjunction is independent of the order between the conjuncts. A double line in an inference rule means that the inference is valid in both directions;[108] with this notation, the bidirectional inference rule

$$\begin{array}{c} A \,\&\, B \text{ true} \\ \hline\hline B \,\&\, A \text{ true} \end{array}$$

is valid. Using this inference rule in the normal way, i.e., from premiss to conclusion, can be seen as an abbreviation of the demonstration

$$\frac{\dfrac{A \,\&\, B \text{ true}}{B \text{ true}} \qquad \dfrac{A \,\&\, B \text{ true}}{A \text{ true}}}{B \,\&\, A \text{ true}},$$

[108]Cf. Martin-Löf, *Intuitionistic Type Theory*, p. 9.

and the inference in the other direction can be seen as an abbreviation of the same demonstration with A and B interchanged. Similarly, the two inference rules making up the bidirectional inference rule

$$\frac{(A \mathbin{\&} B) \mathbin{\&} C \text{ true}}{\overline{A \mathbin{\&} (B \mathbin{\&} C) \text{ true}}}$$

can be motivated by schematic demonstrations.

Next out is disjunction. Remember that a cause of $A \vee B$ consists of a cause of A or a cause of B together with information about which cause it is that is given. This explanation makes the inference rules

$$\frac{A \text{ true} \quad (B : \text{prop})}{A \vee B \text{ true}}$$

and

$$\frac{(A : \text{prop}) \quad B \text{ true}}{A \vee B \text{ true}}$$

self-evident. In these inference rules, the premiss which is needed only to make the inference rule well-formed, i.e., as a presupposition of the conclusion, is put in parentheses. The other logical laws involving disjunction are the Stoic mood *modus tollendo ponens*[109] and proof by dilemma which are expressed by the inference rules

$$\frac{A \vee B \text{ true} \quad {\sim}A \text{ true}}{B \text{ true}}$$

and

$$\frac{A \vee B \text{ true} \quad A \supset C \text{ true} \quad B \supset C \text{ true}}{C \text{ true}}$$

respectively. In these, and similar, inference rules, the leftmost premiss is called the major premiss and the other premisses are called minor premisses. The Stoic mood *modus tollendo ponens* can be justified directly using the meanings of the terms involved; but, as it can also be reduced to more primitive inference rules, this justification is left to the reader at this point. In the disjunctive syllogism, or proof by dilemma, the propositions A and B are called the *horns* of the dilemma and $A \supset C$ and $B \supset C$ are the two lemmata after which this mood of demonstration is named. The justification of proof by dilemma goes as follows: to get a cause of C, first inspect the cause of $A \vee B$; if this consists of a cause of A, invoke the left lemma with this cause of A to get a cause of C; if the cause of $A \vee B$ consists of a cause of B, invoke the right lemma with this cause of B to get a cause of C; in both cases, C has a cause.

Just as *modus tollendo ponens* is justified later, so is the bidirectional inference rule showing that the truth of a disjunction is independent of the order between the two terms involved in it, i.e.,

$$\frac{A \vee B \text{ true}}{\overline{B \vee A \text{ true}}}\,.$$

---

[109] *Modus tollendo ponens*: mood which by denying affirms.

Similarly, the association of parentheses is irrelevant in a disjunction, i.e., the bidirectional inference rule

$$\frac{(A \vee B) \vee C \text{ true}}{\overline{A \vee (B \vee C) \text{ true}}}$$

is valid.

For implication,[110] the most important inference rule is *modus ponendo ponens*:

$$\frac{A \supset B \text{ true} \quad A \text{ true}}{B \text{ true}} .$$

Recall that a cause of $A \supset B$ consists of a method that takes any cause of A into a cause of B; a cause of A is given by the second premiss; combining these ingredients and performing the method results in a cause of B, i.e., the inference rule is evident upon explaining the meanings of the terms involved.

The connective Λ is associated with the logical law *ex falso quodlibet*, i.e., the inference rule

$$\frac{\Lambda \text{ true} \quad (A : \text{prop})}{A \text{ true}} .$$

This inference rule is justified as follows: granted that Λ has a cause c, a cause of A has to be given for each of the possible forms of c; there are no possible forms of c, so this is done by doing nothing; so A has a cause.[111] A perhaps more transparent way of seeing that this inference rule is in fact valid is to compare it to proof by dilemma and *modus ponendo ponens*. The proposition $A \vee B$ is a binary disjunction; a unary disjunction is naturally identified with a proposition A; a nullary disjunction is false and thus identified with Λ. Thus, the propositions $A \vee B$, A, and Λ are in a falling scale. The corresponding inference rules are

$$\frac{A \vee B \text{ true} \quad A \supset C \text{ true} \quad B \supset C \text{ true}}{C \text{ true}}$$

with two minor premisses,

$$\frac{A \text{ true} \quad A \supset C \text{ true}}{C \text{ true}}$$

with one minor premiss, i.e., *modus ponendo ponens* with the premisses reversed, and

$$\frac{\Lambda \text{ true}}{C \text{ true}}$$

with no minor premisses.

[110] I have chosen to take the inference rule *modus ponendo ponens* as meaning determining for implication. In doing so I am faithful to the natural formulation of the BHK interpretation of $A \supset B$, namely that a cause of $A \supset B$ consists of a method taking a cause of A into a cause of B. Another interpretation which, *prima facie*, seems equivalent but which, in fact, is not, is that a cause of $A \supset B$ consists of a cause of B provided that a cause of A is given (this is the interpretation given by Kolmogorov, 'Zur Deutung der intuitionistischen Logik', p. 59, with the only difference that his interpretation is formulated in terms of problems and solutions instead of in terms of propositions and causes). The complete answer to this question is postponed to Ch. V.

[111] This justification will be made more transparent in Ch. V, § 8.

Since negation is defined in terms of implication and *falsum*, there are strictly speaking no inference rules which pertain to negation; instead inference rules involving negation are special cases of other inference rules. For example, the principle of noncontradiction

$$\frac{A \text{ true} \quad {\sim}A \text{ true}}{\Lambda \text{ true}}$$

is a special case of *modus ponendo ponens*. Two of the Stoic moods remain, namely, *modus tollendo tollens*

$$\frac{A \supset B \text{ true} \quad {\sim}B \text{ true}}{{\sim}A \text{ true}}$$

and *modus ponendo tollens*

$$\frac{{\sim}(A \mathbin{\&} B) \text{ true} \quad A \text{ true}}{{\sim}B \text{ true}} .$$

These inference rules can either be justified directly, or demonstrated in terms of more basic inference rules. As the latter demonstrations are given later, the justifications are left to the reader at this point.

There is an important but subtle difference in demonstrating something from known, or accepted, premisses and demonstrating it from premisses which are merely assumed, contingently, as it were. Properly speaking, inferences are made only in the former case, where we pass from something we know to something we get to know. An example will make this clearer. First, think about the letters L, M, P, and F as having the following meanings

$$\begin{cases} \text{L} & = & \text{to be a logician,} \\ \text{M} & = & \text{to be a mathematician,} \\ \text{P} & = & \text{to be a philosopher,} \\ \text{F} & = & \text{to be interested in first principles.} \end{cases}$$

Let it moreover be accepted that a logician is a philosopher or a mathematician, and that a philosopher is interested in first principles, i.e.,

$$\begin{cases} \text{L} \supset (\text{P} \vee \text{M}) \text{ true, and} \\ \text{P} \supset \text{F true} . \end{cases}$$

To get an example of demonstration properly speaking, think about somebody who is a logician but not interested in first principles, i.e., grant that L is true and that ~F is true. It can now be demonstrated that the person you have in mind is in fact a mathematician:

$$\frac{\dfrac{\text{L} \supset (\text{P} \vee \text{M}) \text{ true} \quad \text{L true}}{\text{P} \vee \text{M true}} \qquad \dfrac{\text{P} \supset \text{F true} \quad {\sim}\text{F true}}{{\sim}\text{P true}}}{\text{M true}}$$

On the other hand, if you do not have any particular person in mind but want to demonstrate the proposition

$$(\text{L} \mathbin{\&} {\sim}\text{F}) \supset \text{M},$$

i.e., that *if* somebody is a logician but not interested in first principles *then* he is a mathematician, then demonstration from merely assumed premisses has to be involved.

From Aristotle to Gentzen, logicians took for granted that demonstration from merely assumed premisses follows the same laws as demonstration from accepted premisses.[112] It could have been objected that this practice was unfounded, but I know of no such objection prior to Gentzen. Instead, Gentzen showed how demonstration from assumed premisses is to be understood in terms of demonstration from accepted premisses, and solved the problem before it was formulated.

When demonstrating propositions from assumed premisses the kind of propositions dealt with are *hypothetical*; in this context I understand any proposition of the form $A \supset B$ as hypothetical. Traditionally, the Stoic moods were called hypothetical syllogisms and their major premisses were all called hypothetical propositions, i.e., the propositions $A \vee B$ and $\sim(A \mathbin{\&} B)$ were considered hypothetical, in addition to $A \supset B$;[113] with our definition of negation, the negated conjunction is hypothetical, but the disjunction is not.

The difference, mentioned above, between demonstrating something from accepted premisses and demonstrating something from assumed premisses can now be reformulated as follows: What is the difference between the *validity* of the inference rule

$$\frac{A_1 \text{ true} \quad \cdots \quad A_n \text{ true}}{C \text{ true}}$$

and the *truth* of the implication

$$(A_1 \mathbin{\&} \cdots \mathbin{\&} A_n) \supset C?$$

That an inference rule is valid means that, once the premisses are known, nothing more is called for to come to know the conclusion. That the implication is true means that there is a method which takes a cause of $A_1 \mathbin{\&} \cdots \mathbin{\&} A_n$ into a cause of C. In the inference rule *modus ponendo ponens*, a hypothetical proposition occurs as a premiss, so it seems as if inference is more fundamental than implication. That it has to be so is seen most clearly by Carroll's paradox.[114] If the validity of the inference rule

$$\frac{A \supset B \text{ true} \quad A \text{ true}}{B \text{ true}}$$

was dependent on the truth of the proposition

$$((A \supset B) \mathbin{\&} A) \supset B,$$

we would need the inference rule

$$\frac{((A \supset B) \mathbin{\&} A) \supset B \text{ true} \quad A \supset B \text{ true} \quad A \text{ true}}{B \text{ true}}$$

to reach the conclusion B, but then the validity of this inference rule would be dependent on the truth of the proposition

$$((((A \supset B) \mathbin{\&} A) \supset B) \mathbin{\&} (A \supset B) \mathbin{\&} A) \supset B,$$

---

[112] Cf. Aristotle, *An. Pr.*, Bk. 1, Ch. 1; *An. Post.*, Bk. 1, Ch. 2; Gentzen, 'Untersuchungen über das logische Schließen I & II'; and Sundholm, 'Inference versus Consequence'.

[113] Cf. Boëthius, 'De hypotheticis syllogismis'.

[114] Carroll, 'What the Tortoise said to Achilles'.

etc. *ad infinitum.* The conclusion that B is true would never be reached, as the poor Achilles experienced in Carroll's entertaining description of his paradox.

I broke off the presentation of the intuitionistic account of the laws of propositional logic to explain the difference between demonstration from accepted premisses and demonstration from merely assumed premisses. It would be a terrible blow to logic if its laws could not be justified also in the hypothetical case, but, indeed, they can be.

To avoid confusion between the conjunctions and implications which make up a hypothetical proposition and the conjunctions and implications which are, as it were, active in the inference rules, I will henceforth write

$$A \text{ true } \ (A_1 \text{ true}, \ldots, A_n \text{ true})$$

instead of

$$(A_1 \mathbin{\&} \cdots \mathbin{\&} A_n) \supset A \text{ true} .$$

For the present purposes, there is no need to make a distinction between the meanings of the two ways of expressing a hypothetical proposition, i.e., the bidirectional inference rule

$$\frac{A \text{ true } \ (A_1 \text{ true}, \ldots, A_n \text{ true})}{\overline{(A_1 \mathbin{\&} \cdots \mathbin{\&} A_n) \supset A \text{ true}}}$$

can be taken as defining this new way of writing hypothetical propositions.[115] In the new notation, A_1 true, ..., A_n true, are called the *assumptions* and it is convenient to group them together in a *context* Γ, i.e., to write

$$A \text{ true } \ (\Gamma)$$

instead of

$$A \text{ true } \ (A_1 \text{ true}, \ldots, A_n \text{ true}).$$

For example, the inference rule

$$\frac{A \text{ true} \quad B \text{ true}}{A \mathbin{\&} B \text{ true}}$$

can be generalized to

$$\frac{A \text{ true } \ (\Gamma) \quad B \text{ true } \ (\Gamma)}{A \mathbin{\&} B \text{ true } \ (\Gamma)} \, .$$

If the number of assumptions is zero, the latter inference rule has the same meaning as the former. To justify this inference rule, and indeed any similar inference rule, the method which is a cause of the conclusion has to be defined. It operates by invoking the methods provided by the premisses on the supplied causes for Γ and combines the result using the corresponding inference rule without context, which is already justified. It is clear that all inference rules introduced above can be generalized in this way.

---

[115]Cf. Gentzen, 'Untersuchungen über das logische Schließen I & II', p. 180, n. 2.4. Note that this identification of implication and conditional stands in contrast to distinction upheld between the two by Martin-Löf and Sundholm. The contrast is due to my taking of *modus ponendo ponens* as meaning determining for implication, instead of implication introduction.

Three more inference rules are needed to complete the account of demonstration from assumed premisses. First the inference rule

$$\frac{A_1 : \text{prop} \quad \cdots \quad A_n : \text{prop}}{A_i \text{ true } \ (A_1 \text{ true}, \ldots, A_n \text{ true})}$$

which corresponds to making an assumption (in this inference rule, the number i is between 1 and n). The method which is a cause of the conclusion simply ignores all input apart from the cause of A_i which is given as output.

Next, the inference rule

$$\frac{B \text{ true } \ (\Gamma) \quad A : \text{prop}}{B \text{ true } \ (\Gamma, A \text{ true})}$$

which allows us to add extra superfluous assumptions; this inference rule is called the *weakening rule*. The method which is a cause of the conclusion simply ignores the cause of A and invokes the method which is a cause of the first premiss.[116]

The third inference rule is the rule which allows us to demonstrate an implication by demonstrating the consequent from the assumption of the antecedent.

$$\frac{B \text{ true } \ (\Gamma, A \text{ true})}{A \supset B \text{ true } \ (\Gamma)} \,.$$

The method which is a cause of the conclusion operates as follows: it takes causes γ for the assumptions of Γ as input and gives as output the method which takes a cause a of A into the cause of B given by applying the method provided by the premiss to γ and a.[117]

Having thus explained how demonstration from assumptions works, the laws of logic which were left without justification above can now be justified. First, note that proof by dilemma can be reformulated by

$$\frac{A \vee B \text{ true} \quad C \text{ true } \ (A \text{ true}) \quad C \text{ true } \ (B \text{ true})}{C \text{ true}} \,,$$

which often is more convenient.

The Stoic mood *modus tollendo ponens* is demonstrated below: the premisses are that A and B are propositions, and that the propositions $A \vee B$ and $\sim\! A$ are true. Recall that $A \vee B$ being true presupposes that $A \vee B$ is a proposition, which in turn presupposes that A and B are propositions; thus, the extra premisses of the demonstration are presuppositions of the major premiss.

$$\dfrac{A \vee B \text{ true} \qquad \dfrac{\dfrac{\dfrac{\sim\! A \text{ true}}{\sim\! A \text{ true } \ (A \text{ true})} \quad \dfrac{A : \text{prop}}{A \text{ true } \ (A \text{ true})}}{\Lambda \text{ true } \ (A \text{ true})}}{B \text{ true } \ (A \text{ true})} \qquad \dfrac{B : \text{prop}}{B \text{ true } \ (B \text{ true})}}{B \text{ true}} \,.$$

Note that the demonstration involves a use of *ex falso quodlibet.*

[116] If Γ is empty, the cause of the conclusion is the method which is constantly the cause of B given by the first premiss.

[117] Note that when Γ is empty, the premiss and the conclusion of this inference rule amount to the same.

That the truth of a disjunction is independent of the order of disjuncts is demonstrated as follows:

$$\frac{A \vee B \text{ true} \qquad \dfrac{\dfrac{B : \text{prop}}{A \text{ true } (A \text{ true})}}{B \vee A \text{ true } (A \text{ true})} \qquad \dfrac{\dfrac{B : \text{prop}}{B \text{ true } (B \text{ true})}}{B \vee A \text{ true } (B \text{ true})}}{B \vee A \text{ true}}.$$

The proof that disjunction is associative is not very difficult, but the demonstration becomes so large when written down in full detail that I leave it as an exercise to the reader; it involves two applications of proof by dilemma.

The Stoic mood *modus tollendo tollens* can be seen as a special case of the inference rule

$$\frac{A \supset B \text{ true} \quad B \supset C \text{ true}}{A \supset C \text{ true}},$$

with C taken to be the proposition Λ, because the negation of a proposition A is by definition the proposition $A \supset \Lambda$. Thus, the demonstration

$$\frac{\dfrac{\dfrac{B \supset C \text{ true}}{B \supset C \text{ true } (A \text{ true})} \qquad \dfrac{A \supset B \text{ true}}{B \text{ true } (A \text{ true})}}{C \text{ true } (A \text{ true})}}{A \supset C \text{ true}}$$

is *a fortiori* a demonstration of *modus tollendo tollens*.

Similarly, the Stoic mood *modus ponendo tollens* can be seen as a special case of the inference rule

$$\frac{(A \,\&\, B) \supset C \text{ true} \quad A \text{ true}}{B \supset C \text{ true}},$$

with C taken to be the proposition Λ, since $\sim(A \,\&\, B)$ is by definition $(A \,\&\, B) \supset \Lambda$. Thus, the demonstration

$$\frac{\dfrac{\dfrac{(A \,\&\, B) \supset C \text{ true}}{(A \,\&\, B) \supset C \text{ true } (B \text{ true})} \qquad \dfrac{\dfrac{A \text{ true}}{A \text{ true } (B \text{ true})} \qquad \dfrac{B : \text{prop}}{B \text{ true } (B \text{ true})}}{A \,\&\, B \text{ true } (B \text{ true})}}{C \text{ true } (B \text{ true})}}{B \supset C \text{ true}}$$

is *a fortiori* a demonstration of *modus ponendo tollens*.

Something which is easy to overlook is that the principle that equivalent propositions can be interchanged *salva veritate* cannot be taken for granted, but has to be demonstrated for each form of proposition. That is, the inference rules

$$\frac{A \supset\subset C \text{ true} \quad B \supset\subset D \text{ true}}{(A \,\&\, B) \supset\subset (C \,\&\, D) \text{ true}}, \qquad \frac{A \supset\subset C \text{ true} \quad B \supset\subset D \text{ true}}{(A \vee B) \supset\subset (C \vee D) \text{ true}},$$

and

$$\frac{A \supset\subset C \text{ true} \quad B \supset\subset D \text{ true}}{(A \supset B) \supset\subset (C \supset D) \text{ true}}$$

have to be demonstrated. These demonstrations are left as exercises to the reader.

This completes the justification of the laws of propositional logic under the intuitionistic definition of the notion of proposition.[118]

§ 8. Variables and generality

In the previous sections, I have used letters in the inference rules, standing for various kinds of objects: propositions, numbers, and causes. This use of letters will now be justified.

The Greek geometer Eudoxus, who is said to be the original author of Book V, on proportions, of Euclid's *Elementa*, was perhaps the first to use letters to denote points in the way we are used to in geometry;[119] but the earliest available written account of letters standing for things, in the sense we are investigating, is due to Aristotle: "First, then, let us take a negative universal premiss having the terms A and B. Then if A applies to no B, neither will B apply to any A; ...".[120] The Philosopher's use of A and B as standing for terms, which is introduced without comment, is of great significance to the development of science—logic and mathematics simply could not do without it.

Subsequently, letters were used by many different authors, standing for objects of other kinds. As previously stated, by Euclid in *Elementa*, where one encounters the points A and B, the straight line AB, and the numbers A and B; next, by Boëthius, who uses letters as standing for propositions.[121] For Aristotle, Euclid, and Boëthius, the letter A stands for an indefinite given thing (term, point, number, or proposition). Examples of assertions involving such letters are given by Aristotle's *all A are B*, where A and B are terms, Euclid's *the triangle ABC is equilateral*, where A, B, and C, are points, and Boëthius' *if A is true, then B is true*, where A and B are propositions.

An assertion involving such indefinite given terms, points, or numbers, A and B, is to be interpreted as being true whenever A and B are replaced with *any* particular terms, points, or numbers.[122] I use the word *any* here with reference to Russell's distinction between any and all.[123] Although Russell's explanation of this distinction is unclear, it seems as if he has discovered the conceptual priority of the letters used in inference rules over the variables used in quantification. Thus, I use *any* in connection with inference rules as opposed to *all* which is used in connection with the universal quantifier. Wittgenstein's view is that an inference rule can only show, or exhibit, generality, not express

[118]This way of manipulating propositions, exemplified above, soon becomes rather cumbersome, as the assumptions are repeated on each line of the demonstration. A more economical way of expression is used in *natural deduction* (introduced by Gentzen, op. cit.). There are several good textbooks on the subject, e.g., van Dalen, *Logic and Structure*; cf. Prawitz's monograph *Natural Deduction*.

[119]Kneale and Kneale, *The Development of Logic*, p. 61.

[120]Aristotle, *An. Pr.*, Bk. 1, Ch. 2. The universal negative is *no A is B* which implies *no B is A* because, as we would say, A and B are disjoint.

[121]Boëthius, 'De hypotheticis syllogismis', passim.

[122]Concerning Aristotle's use of letters, and the way that they may be replaced, cf. Aristotle, *An. Pr.*, 25a14 and 26a8.

[123]Russell, 'Mathematical Logic as Based on the Theory of Types', § 2.

it, and this fits well with the above.[124] It cannot be demonstrated that an inference is an instance of an inference rule, though the generality of the inference rule is still *exhibited* by the use of schematic letters.

Next, it must be decided what to call such letters standing for indefinite given things. We have at least the following proposals: Leibniz's parameter,[125] Russell's real variable,[126] Post's operational variable,[127] Quine's schematic letter,[128] placeholder,[129] metavariable,[130] dynamic variable,[131] and, finally, Schütte's Mitteilungszeichnen.[132] Of course, there are different nuances to the different terminologies. Subsequently I will prefer Quine's term schematic letter. Sometimes I also use Leibniz's term parameter for mathematical objects, like numbers, in conformity with mathematical practice.

In the above presentation of propositional logic, I have only used parameters and schematic letters. What about the ordinary variables ubiquitous is mathematics? The first distinction akin to the distinction between parameters and variables was made by Diophantus in his book on arithmetic, which deals with the solution of equations.[133] It does not make sense to formulate an equation with constants if these constants are assumed to be known, as is the case with parameters. Here I take given to be equivalent with known. For example, consider the equation $100 = 2\varsigma + 40$. If ς is a known number, then this is just an assertion, or a proposition, depending on how the equality sign is interpreted. But to ask for which values of ς the equation holds, it must be assumed that ς is unknown. Diophantus was aware of this problem and, to solve it, he distinguishes the known from the unknown. The sought number is represented by the letter ς. Here I take sought to be equivalent with unknown. He also introduces a sign for the arithmetical unit, viz., $\overset{\circ}{M}$, and employs it in front of known numbers.[134] This enables him to write $\varsigma\bar{\beta}\ \overset{\circ}{M}\,\bar{\mu}$ for $2\varsigma + 40$ and solve the equation $100 = 2\varsigma + 40$.[135] Later on Vieta established a little known typographical distinction between known and unknown quantities in *Artem*

---

[124] Cf. Goodstein, 'Function Theory in an Axiom-Free Equation Calculus', p. 434.
[125] Leibniz, *Mathematische Schriften*, p. 268.
[126] Russell, 'Mathematical Logic as Based on the Theory of Types', § 2. Concerning Russell's choice of terminology: certainly A and B are variables in the sense that they admit various replacements, but not in the sense that they are assumed to vary throughout the argument. In fact, they are constant in exactly the sense that they are *not* supposed to vary throughout the argument. The ordinary variables were called *apparent* by Whitehead and Russell who attribute this choice of terminology to Peano (*Principia Mathematica*, Intro., p. 52).
[127] Post, 'Formal Reductions of the General Combinatorial Decision Problem'.
[128] Quine, *Methods of Logic*, p. 235.
[129] E.g., Magnusson, 'The Implementation of ALF'.
[130] E.g., Sato et al., 'Calculi of Meta-variables'.
[131] Moreau, 'A Syntactic Theory of Dynamic Binding'.
[132] Schütte, *Beweistheorie*, p. 9.
[133] Diophantus, *Arithmeticorum Liber Sex et de Polygonis Numeris Liber*, p. 7.
[134] Unit is μονάς in Greek and *unitas* in Latin.
[135] The letter ς is the variant of σ used at the end of a word, *final sigma*. Diophantus' choice of this letter is perhaps due to the fact that all normal letters of the Greek alphabet already were used to represent numerals. For example, $\bar{\beta}$ and $\bar{\mu}$ are the Greek numerals 2 and 40 respectively. Another explanation is given by Klein, *Greek Mathematical Thought and the Origin of Algebra*, p. 146, referring to Heath, q.v., who conjectures that ς is a ligature for αρ, which in its turn is an abbreviation of ἀριθμός, i.e., number.

Analyticam Isagoge.[136] He proposed the use of vowels for unknown quantities and consonants for known quantities, i.e., the ς of Diophantus is translated by an a by Vieta. Descartes modified Vieta's convention and used letters from the end of the alphabet for unknown quantities and letters from the beginning of the alphabet for known quantities,[137] and this convention still remains in force.

The word variable was first used by Leibniz,[138] who contrasts variables with constants and parameters,[139] where constant is to be taken in a narrow sense, i.e., as a meaningful categorem or a *name* in the usual sense of the word. To understand the notion of variable better, consider the following list of opposites:

constant	vs.	variable
definite	vs.	indefinite
given	vs.	sought
known	vs.	unknown
determinate	vs.	indeterminate

For each pair of opposites, it must be determined what genus they belong to. For the first pair of opposites, constant vs. variable, I have already indicated that the genus is mathematical categorem. The opposites definite and indefinite apply to constants.[140] Given and sought apply to indefinite constants, e.g., in $ax^2 + bx + c = 0$, a, b, and c are given and x is sought. Known and unknown have wider extension, viz., mathematical categorem, but, for indefinite constants, known agrees with given and sought agrees with unknown. The above distinctions motivate the classification given in Table 3.

For schematic letters, a crucial observation is that for an assertion or inference rule involving them to be valid, it is *not* necessary to be able to enumerate, or otherwise exactly specify, the range of the schematic letter; however, it *is* necessary to know which logical category the schematic letter belongs to.[141]

It remains to account for the normal use of the words determinate and indeterminate. The last pair of opposites, determinate and indeterminate, also apply to mathematical categorems. A mathematical categorem is called determinate if it is taken as referring to something specific. With respect to the threefold correspondence in Figure 1, the triangle must be completed, but the object, i.e., the reference, may still be in shadow, i.e., to be found. Two other common qualifiers are fixed and arbitrary, and often in combination.

---

[136]Vieta, *Artem Analyticam Isagoge seu Algebra Nova*, Ch. 5, § 5. This book was first published 1591.

[137]Descartes, *Discours de la Méthode.*

[138]Kline, *Mathematical Thought from Ancient to Modern Times*, p. 340.

[139]Leibniz, *Mathematische Schriften*, p. 268.

[140]An analogue to the distinction between definite and indefinite constant quantities is found in certain programming languages, e.g., in C, viz., the distinction between a definition and a declaration, or typing. The declaration makes the constant known, but left indefinite, and the definition makes the constant into a name. For example, "int f(int x) ;" is a declaration, and "int f(int x){return $x * x$;}" is a definition.

[141]The first of these insights is related to Aristotle's solution to the dilemma reached in Plato's dialog Meno (*An. Post.*, Bk. 1, Ch. 1, c. 71a29). However, a problem with Aristotle's term logic is that it does not make it clear which logical category the schematic letters range over.

With respect to its *referent*, a categorem, i.e., an atomic expression, is either:

- a *constant*, i.e., its referent is *fixed* throughout the discourse:
  - a constant is either a *definite* constant, i.e., a *name*, or *categorem* with full meaning; or
  - an *indefinite* constant, which is either:
    - *given* or *known*, i.e., a *parameter* or a *schematic letter*, which is conceived as a given object of a certain logical category; or
    - *sought* or *unknown*, i.e., the kind of quantity used in equation solving;
- a *variable*, i.e., its referent is conceived as varying.

Table 3. Classification of mathematical categorems into variables and constants, definite and indefinite, and given and sought.

This mystery is solved by identifying fixed with constant and arbitrary with indeterminate, so a fixed and arbitrary quantity is a schematic letter. Combining known–unknown with determinate–indeterminate, we get four possibilities. Consider the following table:

	determinate	indeterminate
known	name	parameter
unknown	sought	variable

From the above, it is clear that name, parameter, and variable are in the right position. By sought, is to be understood Diophantus' ς, i.e., the thing sought in an equation, or an indefinite and unknown constant. In a certain sense, it is correct to say that, that which is sought is unknown but determinate. For example, in the case of a system of linear equations, which indeed is called determinate exactly when it has a unique solution, and under- or over-determined if it has many or no solutions. In other cases however, it is slightly incorrect to say that what is sought is determinate. For example in the case of polynomial equations, where the solution is only determined to one of a finite number, or in the case of posing an equation without knowing if there is a solution. To understand in what sense the x and y in an equation, or system of equations, are unknown but determinate, consider a quiz like the following: I have two sons—together they are 33 years old, and one is three years older than the other. How old are my sons? A mathematician automatically writes

$$\begin{cases} x+y &= 33, \\ x-y &= 3, \end{cases}$$

and solves the system for x and y, giving $x = 18$ and $y = 15$. The terms x and y, standing for the ages of the two sons, were certainly determined all along, but became known only by means of the computations.

Concerning the notation used for variables, it seems as if Leibniz, Newton, and Euler, adopted, or modified, Descartes' convention, and used z, y, x, etc. for variables. The reason for this is probably that in an equation like $ax+b=0$, the x has the dual rôle of being sought in the equation and being a variable in the function $ax + b$. Another typographical convention which seems to be gaining force is to use an upright font for definite constants, like sin and log, and to italicize indefinite constants and variables. There are several exceptions to this rule, e.g., that the base e of the natural logarithm usually is printed in italics.

Of course, variables are used also in intuitionistic type theory, I have only not introduced them yet. Variables in logic are related to one of the largest contributions to logic since Aristotle, namely, Frege's idea of replacing the classical square of opposition with two quantifiers for forming universal and particular propositions.[142] When bringing the quantifiers into logic, variables almost inevitably come with them. Variables and quantifiers will be treated of in full detail in Chapter V.

The key to Frege's insight is to recognize that the word *is* is used in two very different senses in the two sentences *zero is a number* and *zero is not equal to one*. This distinction should be compared to the Aristotelian distinction between essential and accidental predication.[143] The translations of these two sentences into intuitionistic type theory are

$$0 : \mathrm{number}$$

and

$$\sim(0 \ \mathrm{eq}\ \mathrm{s}(0)) \ \mathrm{true},$$

Note that they have different form: one is complete and the other is incomplete. What is it that 0 *is* in the second of these two assertions? Frege's answer to this question is that 0 is

$$\sim(x \ \mathrm{eq}\ \mathrm{s}(0)),$$

viewed as a function of x, and this is how variables came into logic.[144] With Frege, I will call such a thing a *propositional function*. Another example is

$$x < \mathrm{s}(\mathrm{s}(0)),$$

i.e., x *is less than two*. Let these examples suffice for the moment. I write $P(x)$ for such a propositional function and $P(a)$ for the proposition which is the result of replacing all relevant occurrences of x by a. The general pattern of understanding accidental predication can be described as follows. Translate the original sentence a *is* Q into a proposition B so that a *is* Q is tantamount to B being true. I will assume that a is translated to c. Replace the parts of B that correspond to c with an x.[145] This gives a propositional function $P(x)$ such that $P(c)$ is equal to B. The translation of a *is* Q is now tantamount to $P(c)$ being true, and it is natural to say that P is the translation of Q, e.g., we say that the propositional function $\sim(x \ \mathrm{eq}\ 0)$ corresponds to the predicate

[142] Frege's *Begriffsschrift* is the source of this idea, but it has gone through several refinements before its present form. Quantifiers were also discovered independently by Peirce, 'On the Algebra of Logic: a contribution to the philosophy of notation', p. 194.

[143] Cf. *Metaph.*, Bk. 5, Ch. 7; and *Cat.*, passim.

[144] Cf. Frege, 'Function and Concept'.

[145] Note that this does not have to be all occurrences of c, as the above example shows: c is 0, while $P(x)$ is not $\sim(x \ \mathrm{eq}\ \mathrm{s}(x))$ but $\sim(x \ \mathrm{eq}\ \mathrm{s}(0))$.

not equal to zero, and that the propositional function $x < \mathrm{s}(\mathrm{s}(0))$ corresponds to the predicate *less than two*.

Note how flexible natural language is with respect to different senses of the word *is*. The essential predication 0 : number can be read *zero is a number*, the assertion of the propositional equality (2 + 2) eq 4 true can be read *two plus two is four*,[146] the accidental predication prime(3) true can be read *three is a prime*, and the accidental predication $3 < 4$ true can be read *three is less than four*.

§ 9. Division of definitions

First, a distinction is to be made between *nominal* and *real* definitions. Traditionally, the distinction between nominal and real definitions is explained by saying that a nominal definition defines what a word or phrase *means* whereas a real definition defines what something *is*.[147] This distinction requires some clarification because, at least for beings of reason, there is no real difference between giving the meaning of a word or phrase and explaining what it is, i.e., what the word or phrase refers to.[148] On the other hand, I think that the distinction can be maintained if properly clarified. Thus, a nominal definition defines what a word or phrase, of a certain logical category, means *by reducing it to an already understood expression in the same logical category.* Thus, a nominal definition is nothing but an abbreviatory definition. It is now clear that not all definitions can be nominal, i.e., the most primitive forms of expression have to be given definitions which are not nominal, if they are to be defined at all.

It is not as easy to explain what a real definition is. One way out is to make use of the distinction between object language and metalanguage,[149] in which case a real definition of a word or phrase in the object language consists in explaining its meaning in metalanguage. I will take this distinction as my starting point. A distinction was made above between *nonsense* and *absurdity*. It should be clear that if a real definition is understood in the way explained above, it endows the word or phrase defined with *sense*, though it may still be absurd. For example, when we make a definition like "a *group* is a set with a binary operation satisfying..." there is *prima facie* no guarantee that there are groups; this is clear because this definition has the same *form* as the definition "an *infinite number* is a number greater than or equal to any number", and there are no infinite numbers. It may not always be apparent whether a definition is absurd or not, e.g., if "a *chimera* is a finite non-abelian group of prime order", then it takes some group theory to show that chimeras are absurd.

Recall the distinctions made above between essential and accidental predication and between complete and incomplete assertions. The kind of definition

[146]The definitional equality 2 + 2 = 4 : number can also be read using *is*, but it is not so natural: *two plus two is by definition the number four*, or something like that.

[147]Gredt, *Elem. Phil.* Ch. 2, § 4, nn. 32–35.

[148]That is, the distinction is as small as the distinction, made on p. 21, between objective concept and formal object.

[149]Taken from Curry, *Foundations of Mathematical Logic*, Ch. 2, § A.

exemplified above, which I will call a descriptive definition, always defines an accident of something, i.e., if G is a set and f is a binary operation on G, then group(G, f) and chimera(G, f) can be made into perfectly good propositions, and, if n is a number, then infinite(n) can also be made into a proposition.

When defining a complete assertion, i.e., an essential form of predication, the situation is different. For example, one may try to define zero and the successor of a number by descriptive definitions like "zero is the least number" and "the successor of a number is the number following immediately after it", facing the same problem of existence as above; if, on the other hand, these definitions are reformulated as "the form of assertion a : number holds when a has one of the forms 0 or s(b), where b : number", then there is no question of existence for zero and successor. A logical category can always be introduced together with the *forms* of the objects falling under it without there being any question of existence for these forms of objects. They have, as it were, completely shallow meaning, i.e., their only meaning consists in their being objects of the logical category in question.[150]

For real beings, descriptive definitions are unproblematic since the thing defined is guaranteed to exist. On the other hand, when they are employed for beings of reason (*ens rationis*), they are very problematic, since they presuppose a kind of Platonic universe of ideas. I will call the definition of a complete form of assertion a *meaning explanation.* That is, in intuitionistic type theory, descriptive definitions are avoided and replaced by meaning explanations, thereby avoiding the existence problem.

The classical division of real definitions may be of some help in the division of meaning explanations. First we have definitions by various kinds of causes: by efficient cause, i.e., a genetic definition, e.g., "a *circle* is what comes about by moving a point along a trajectory equidistant from a certain fixed point"; by final cause, e.g., "a *screwdriver* is a tool for turning screws into or out of their places", "a *clock* is a machine for the measurement of time", or "a *chair* is a piece of furniture for one man to sit on". These definitions could also be called instrumental definitions. Next we have the metaphysical definition, e.g., "a *man* is a rational animal", and the physical definition, by matter and form. Finally we have the etymological definition, e.g., "a *philosopher* is a lover of wisdom", and the descriptive definition, e.g., "a *man* is a land-living featherless animal walking upright on two legs".[151]

In mathematics we are used to inductive definitions, like the above definition of the numbers. These could be likened to the physical definitions mentioned above, if we instead of matter and form read parts and form. This identification of matter and parts is justified since, after Ockham, the copula in an assertion of the form S is A, was called the form and the terms the matter,[152] while I call the terms the parts of the assertion. Moreover, the mathematical coinductive definitions are similar to definitions by final cause and the definition of the form of assertion A : prop, given above, could be

---

[150]More precisely, taken formally they have shallow meaning, but there can of course be more meaning attached to them outside the formal language, as is the case with zero.

[151]These examples are taken from the *O.E.D.* and from Gredt, *Elem. Phil.* Ch. 2, § 4, n. 33.

[152]Ibid., Ch. 2, § 6, n. 39.

labelled a genetic definition as it describes how a proposition A comes about, viz., by laying down what counts as a cause of it.

CHAPTER III

The Notion of Set

THE NOTION OF SET is central to modern foundations of mathematics, regardless of school. In fact, the position taken on this notion highlights major differences between the schools, but remains central to all of them. The history of the definition of this notion is the history of how universals made into objects of thought are brought into the language of logic proper, i.e., brought from the metalanguage to the object language.

The first section of this chapter gives an historical survey of set-like notions. Next follows a note on set-theoretical notation. In the third section I attempt to capture the naïve notion of set by a descriptive definition. This naïve notion is then expounded in the following section where canonical sets and elements are treated in full detail. This, together with the treatment of noncanonical sets and elements in Chapter IV, Section 4, gives the exact definition of the notion of set as it is understood in intuitionistic type theory. In the final two sections of this chapter, I introduce some of the most important sets of intuitionistic type theory.

§ 1. A history of set-like notions

Here I will analyse the logical notions *set*, *type*, *universe of discourse*, *class*, *system*, and *species*, employed by a number of authors, from Bolzano to the present. To complicate matters, different authors have given different names to essentially the same notion and sometimes the same name to different notions. The most important names are the following:

universe, associated with De Morgan, Venn, and Carroll;

class, associated with Mill, Boole, and Peano;

set, associated with Bolzano, Cantor, and Zermelo;

system, associated with Dedekind;

type, associated with Russell, Whitehead, Church, et al.; and

species, associated with Brouwer, Heyting, Kreisel, and Troelstra.

There are two important distinctions to be made. First, with respect to the meanings of the terms, one should make a distinction between *intensional* and *extensional* notions of set, class, etc. If a concept is taken as standing for the totality denoted, then it is taken extensionally; whereas if the identity of the concept itself is also taken into account, then it is taken intensionally. For example, the concepts *prime number* and *irreducible number* are distinct as concepts, but a theorem of arithmetic says that they have the same extension, i.e., that every prime number is irreducible and conversely. Thus, they are

J.G. Granström, *Treatise on Intuitionistic Type Theory*, Logic, Epistemology, and the Unity of Science 22, DOI 10.1007/978-94-007-1736-7_3,

	Essential	Accidental
Intensional	Set : Martin-Löf. Universe of discourse : De Morgan, Venn, Carroll. Type : Whitehead, Russell.	Species : Brouwer, Heyting, Kreisel, Troelstra.
Extensional	Set : Bishop. Setoid : Hofmann.	Set : Bolzano, Cantor, Zermelo. System : Dedekind. Class : Boole, Peano.

Table 4. The four different notions of set and an approximate classification of the various definitions.

extensionally equal but intensionally distinct concepts.[1] A notion of set, class, etc. is called *extensional* if two sets A and B are considered equal whenever the corresponding concepts are extensionally equal, i.e., if every element of A is an element of B and conversely. If a notion of set, class, etc. is not extensional, then it is called *intensional*. Thus, for a notion of set, class, etc. to be intensional, equality between sets has to mean something more than mere coincidence of the extensions of the concepts.

The second important distinction is whether each element has a unique, or at least primary, set, class, etc. to which it belongs, or the same element can belong to several sets indiscriminately. In the first case, the notion will be called essential and in the second case it will be called accidental.[2] Thus, the outstanding feature of an essential notion of set, class, etc. is that each element has a natural habitat.[3] Take for example the number 3, which first and foremost, essentially, is a number so that, at least, 3 is an element of the set of numbers. One could also allow for the formation of sets like the set of primes and the set of odd numbers, and an accidental notion of set, class, etc. arises if they are viewed as being on a par with the set of numbers. The distinction between essential and accidental notions could be refined into a spectrum by putting the absolute essentialism of intuitionistic type theory at one end of the spectrum and the theories with a universe of everything, like Zermelo-Fraenkel set theory, at the other end of the spectrum; but what I attempt is only a broad classification.

The various intensional notions are investigated first. The notion of the universe of a proposition was introduced by De Morgan and was called the universe of discourse by Venn.[4] Carroll defines the universe of discourse as follows: "The genus, of which the two terms of a proposition are species, is called its *universe of discourse*."[5] Carroll's motivation for the introduction of the universe of discourse was to make the Aristotelian doctrine of syllogisms

[1]Cf. Troelstra, *Principles of Intuitionism*, § 4.2.
[2]The same distinction is made on p. 48 of this book.
[3]Cf. the notion of *principal type* used in computer science.
[4]De Morgan, 'On the Structure of the Syllogism', p. 380; and Venn, *Symbolic Logic*, p. 62.
[5]Carroll, *Symbolic Logic*, p. 70.

live up to modern standards of logical rigor.[6] If Carroll's definition is applied to a proposition like *two and two is four*, it becomes clear that the universe of discourse is similar to a genus in Aristotle's noetic, in this case, number or discrete quantity. The notion of universe of discourse is of intensional nature, and is similar to the notion which Russell calls a type—a type being defined as the range of significance of a propositional function.[7] In fact, Russell was accused of inventing a new name, viz., type, for an existing notion, viz., universe of discourse.[8]

Cantor's second definition of the notion of set, conceived in 1895, before the advent of type theory, reads as follows:

"By a 'set' we understand every collection M of definite well-distinguished objects m of our intuition or thought (which are called the 'elements' of M) to a whole."[9]

This definition seems to be of an intensional notion, but the subsequent use of it makes it plausible that it is intended merely to be a reformulation of his earlier extensional definition.[10] Compare the above definition to Martin-Löf's definition of the notion of set in intuitionistic type theory:

"A set A is defined by prescribing how a canonical element of A is formed as well as how two equal canonical elements of A are formed... There is no limitation on the prescription defining a set, except that equality between canonical elements must always be defined in such a way as to be reflexive, symmetric and transitive."[11]

One could perhaps take the liberty to read Cantor's definite as referring to the ways of forming objects and well-distinguished as referring to the ways of forming equal objects and, in this case, Cantor's second definition is a precursor to Martin-Löf's definition, though, admittedly, this explanation is a little far-fetched.

The notion of data type in programming languages is also of intensional nature. One could take the notion of data type, as used in computer programming, to be synonymous with the notion of set, as used in intuitionistic type theory.[12] The advantage of using intensional sets when modelling computer programs is that extensionally equal functions are not identified as they are in extensional set theory.

The above intensional notions are also classified as essential, except, perhaps, for Cantor's definition. To find an intensional and accidental notion, the first obvious choice is the notion of property or predicate. If by *set* is understood

---

[6]Cf. Aristotle, *An. Post.*, Bk. 1, Ch. 7, where the universe of discourse is anticipated.

[7]First defined in *The Principles of Mathematics*, § 497 and further developed in 'Mathematical Logic as Based on the Theory of Types', p. 236.

[8]Brown, 'The Logic of Mr. Russell'.

[9]Author's translation of Cantor, 'Beiträge zur Begründung der transfiniten Mengenlehre', § 1: "Unter einer ‚Menge' verstehen wir jede Zusammenfassung M von bestimmten wohlunterschiedenen Objekten m unserer Anschauung oder unseres Denkens (welche die ‚Elemente' von M genannt werden) zu einem Ganzen."

[10]Quoted on p. 57 of this book.

[11]Martin-Löf, *Intuitionistic Type Theory*, p. 8.

[12]Ibid., p. 23. Cf. also Martin-Löf, 'Constructive mathematics and computer programming' and Cardelli and Wegner, 'On Understanding Types, Data Abstraction, and Polymorphism'.

a predicate on, say, the natural numbers then, as seen above, this is an intensional notion; moreover, since all predicates are on a par, it is also accidental. The notion of set which occurs in the realizability model of type theory, where a set is interpreted as a predicate on the natural numbers, and an element of a set as a natural number satisfying the predicate, is an elaboration of this idea. In this context Brouwer's notion of species should also be mentioned; the word species is here used in a technical sense.[13] "Roughly speaking, species are properties which are in turn considered as mathematical objects (entities)."[14] A species is sometimes considered intensionally and sometimes extensionally, but it is always accidental, like the notion of predicate discussed above.

Our next definition, due to Bolzano, is interesting mainly because it contains the first use of the word *set*, or *Menge* in German, in its technical sense. Here I have classified it as an extensional and accidental notion, since it is a precursor to Cantor's definitions; but, admittedly, the definition is rather vague, so this classification is open to debate.

"An aggregate whose basic conception renders the arrangement of its members a matter of indifference (and whose permutation therefore produces no essential change from the current point of view), I shall call a *set*, and a set whose members are considered as *individuals* of a stated species A, that is, as objects subsumable under the concept A, is called a *multitude* of A."[15]

Although Bolzano's definition is of an early date, it was not very influential. Instead, Bolzano's ideas on sets and the infinite live on in the form given to them by Cantor and Dedekind, see below.

Let us now consider the extensional notions. The Brouwerian notion of species is the likely source of Bishop's notion of set, which in turn is the source of Martin-Löf's notion of set.[16] The difference is that while Brouwer's notion is primarily intensional, Bishop's notion is extensional. Here is Bishop's definition quoted in full:

"A set is not an entity which has an ideal existence: a set exists only when it has been defined. To define a set we prescribe, at least implicitly, what we (the constructing intelligence) must do in order to construct an element of the set, and what we must do to show that two elements of the set are equal."[17]

Bishop's subsequent use of this notion makes it clear that it is intended as an extensional notion. However, it is certainly an intuitionistic notion of set as opposed to most of the extensional notions of set, type, class, etc. of an earlier date. The notion of setoid, defined as an intensional set together

[13]Cf. Heyting, *Intuitionism: An Introduction*, § 3.2.1.
[14]Troelstra, *Principles of Intuitionism*, § 4.1.
[15]Bolzano, *Paradoxes of the Infinite*, § 4: "Einen Inbegriff, der wir einem solchen Begriffe unterstellen, bei dem die Anordnung seiner Teile gleichgültig ist (an dem sich also nichts für uns Wesentliches ändert, wenn sich bloß diese ändert), nenne ich eine *Menge*; und eine Menge, deren Teile alle als *Einheiten* einer gewissen Art A, d.h. als Gegenstände, die dem Begriffe A unterstehen, betrachtet werden, heißt eine *Vielheit* von A." Note that the German word *Art* was translated by Steele as *species*; both words should be understood in the most general possible sense.
[16]Martin-Löf, 'An intuitionistic theory of types', p. 76.
[17]Bishop and Bridges, *Constructive Analysis*, Ch. 1, p. 5.

with an equivalence relation, is very similar to Bishop's notion of set.[18] I have classified the notion of setoid as an extensional notion since two setoids are normally considered equal when they have the same underlying intensional set and their equivalence relations are equivalent.[19]

Next we have the notion of class, which is usually defined as the extension of a concept. Boole used the word class for the extensional notion which he defines as follows:

"The universe of conceivable objects is represented by 1 or unity. This I assume as the primary and subject conception. All subordinate conceptions of class are understood to be formed from it by limitation, according to the following scheme."[20]

There is no need to investigate Boole's scheme for forming classes to recognize that his definition contains an idea seminal to modern set theory, viz., the "universe of conceivable objects". This supreme genus of everything inspired a range of notions, like Peano's notion of class, Cantor's notion of set, and Dedekind's notion of system.[21] One characteristic feature of these theories is that classes, sets, and systems, are formed by separation; that is, given a propositional function $P(x)$ on the universe of conceivable objects, one can form the set which contains as elements those objects of the universe of conceivable objects which satisfy the predicate $P(x)$, typically written

$$\{x \mid P(x) \text{ true}\},$$

and two such sets are considered equal if they agree in extension. That is, two sets defined by different predicates, say $P(x)$ and $Q(x)$, are considered equal if, for all x in the universe of conceivable objects, $P(x)$ is true if an only if $Q(x)$ is true. It is a too naïve a treatment of this universe of conceivable objects which is shown to be self-contradictory by Russell's paradox.[22] Aristotle argued, albeit inconclusively, that there cannot be a supreme genus of being and,[23] as Bocheński sarcastically puts it: "The same result was reached again in 1908, after Aristotle's doctrine had been forgotten."[24]

The definitions due to Cantor and Dedekind will complete this survey of the period of naïve set theory, which I delimit in time from Boole's definition to Russell's paradox. Cantor's first definition of the notion of set reads as follows:

"A manifold (an aggregate, a set) of elements, which belong to some arbitrary conceptual sphere, I call *well-defined*, if on the basis of its definition and in consequence of the logical principle of excluded middle it must be recognized as *internally determined*, *both* whether an arbitrary object belonging to the same conceptual sphere belongs to the manifold as an element or not, *and also*

---

[18]Cf. Hofmann, 'Extensional concepts in intensional type theory'.

[19]This makes *setoid* an extensional notion if we take the liberty to view the extension of a setoid, not as the extension of the underlying intensional set, but as the class of equivalence classes.

[20]Boole, 'The Calculus of Logic', p. 184.

[21]For Peano's notion, vid. *Arithmetices Principia Nova Methodo Exposita*, Log. Not., n. 4. Cantor's and Dedekind's notions are quoted below.

[22]*The Principles of Mathematics*, § 78 and § 500, cf. the 1908 paper 'Mathematical Logic as Based on the Theory of Types'.

[23]*Metaph.*, Bk. 3, Ch. 3.

[24]*Ancient Formal Logic*, § 6C.

whether two objects belonging to the set, in spite of formal differences in the manner in which they are given are equal or not."[25]

Although Dedekind's definition of the notion of system is of a later date than Bolzano's and Cantor's definitions, it was conceived independently, as Dedekind himself writes in the preface to the second edition of his essay on the nature and meaning of number. Dedekind's definition of the notion of system reads as follows:

"It very frequently happens that different things, a, b, c,... for some reason can be considered from a common point of view, can be associated in the mind, and we say that they form a *system* S".[26]

Dedekind then continues by stating that a system is completely determined when, with respect to every thing, it is determined whether it is an element of S or not and that two systems are to be considered equal when every element of the first is also an element of the second and conversely.

Because of Russell's paradox, the formulation of set theory which eventually became standard is essentially due to Zermelo. Therefore it is of interest to look at his definition:

"Between the things of the domain $\mathcal{B}$ certain "*ground relations*" of the form $a \,\varepsilon\, b$ hold. If for two things a, b, the relation $a \,\varepsilon\, b$ holds, then we say that "a is an *element* of the set b", or "b contains a as element", or "has the element a". A thing b, which contains another thing a as element, can always be called a *set*, but also only then—with only one exception (Axiom II)."[27]

Here a real definition of the notion of set is given up entirely and instead the containment relation ε is taken as primitive. In such an axiomatic formulation of set theory, it is difficult to say exactly what it *means* for something to be a set. By set, one is free to understand anything, as long as it satisfies the axioms; or even more boldly, the notion of set is *defined* by the axioms. The path taken below, in attempting a real definition of the notion of set, is based on another philosophy, viz., that the definition comes first and that the axioms have to be valid in virtue of the definition.

---

[25]Author's translation of Cantor, 'Über unendliche, lineare Punktmannichfaltigkeiten', p. 114, sq.: "Eine Mannigfaltigkeit (ein Inbegriff, eine Menge) von Elementen, die irgendwelcher Begriffssphäre angehören, nenne ich *wohldefiniert*, wenn auf Grund ihrer Definition und in Folge des logischen Prinzips vom ausgeschlossenen Dritten es als *intern bestimmt* angesehen werden muss, *sowohl* ob irgendein derselben Begriffssphäre angehöriges Objekt zu der gedachten Mannigfaltigkeit als Element gehört oder nicht, *wie auch* ob zwei zur Menge gehörige Objekte, trotz formaler Unterschiede in der Art des Gegebenseins einander gleich sind order nicht."

[26]Dedekind, 'The Nature and Meaning of Numbers', § 1: "Es kommt sehr häufig vor, daß verschiedene Dinge a, b, c... aus irgendeiner Veranlassung unter einem gemeinsamen Gesichtspunkte aufgefaßt, im Geiste zusammengestellt werden, und man sagt dann, daß sie ein *System* S bilden".

[27]Author's translation of Zermelo, 'Untersuchungen über die Grundlagen der Mengenlehre', p. 262: "Zwischen den Dingen des Bereiches $\mathcal{B}$ bestehen gewisse „*Grundbeziehungen*" der Form $a \,\varepsilon\, b$. Gilt für zwei Dinge a, b die Beziehung $a \,\varepsilon\, b$, so sagen wir „a sei *Element* der Menge b" oder „b enthalte a als Element" oder „besitze das Element a". Ein ding b, welches ein anderes a als Element enthält, kann immer als eine *Menge* bezeichnet werden, aber auch nur dann—mit einer einzigen Ausname (Axiom II)." For completeness, it should be added that Zermelo's axiom II says that there is an empty set.

§ 2. Set-theoretical notation

As an appendix to this survey of the classical definitions of the notion of set, something has to be said about set-theoretical notation. The use of the epsilon sign for set membership is due to Peano: "The sign ε stands for *is*. Accordingly $a\,\varepsilon\,b$ is read *a is a certain b*."[28] As seen above, this symbol was adopted by Zermelo, and a stylized version, $\in$, of this notation became standard in set theory due to the influence of *Principia Mathematica* and was also used by Martin-Löf in early formulations of intuitionistic type theory.[29]

However, since this symbol is not available on a standard keyboard, it was never accepted in computer science, where instead a colon is used. The colon notation is commonly attributed to de Bruijn and certainly it is present in later writings on Automath,[30] but the colon is also used for type annotations in the programming languages Pascal and ML, conceived in the 70's, and it would be interesting to know who inspired whom. It might also be that the colon notation draws on legacy from the calculus of relations,[31] or from the standard mathematical notation $f : A \to B$.

In the present formulation of intuitionistic type theory, both the epsilon and the colon are used as copula, but with a slight difference in meaning, see below.

§ 3. Making universal concepts into objects of thought

It is now time to propose a new definition of the notion of set which captures the essence of the classical definitions. With respect to the preceding distinctions between intensional and extensional and between essential and accidental, the intensional and essential notion is the most fundamental, since the other notions can, more or less faithfully, be interpreted in terms of it.[32] Therefore I propose the following descriptive definition of the intensional and essential notion of set.

Definition 3. *A* set *is a universal concept, a first intention, collected to a whole, i.e., substantiated, where the objects falling under it, called* elements *of the set, are considered as individuals, and for which the definitional equality between objects satisfies the four requirements: that a definitum is always equal to its definiens, that two objects of the same form are equal if their parts are equal, that any object is equal to itself, and that two objects which equal a third object are equal to one another.*

---

[28]Author's translation of Peano, *Arithmetices Principia Nova Methodo Exposita*, Log. Not., n. 4, p. 27: "Signum ε significat *est*. Ita $a\,\varepsilon\,b$ legitur *a est quoddam b*." The choice of epsilon is, most likely, due to the Greek counterpart ἐστί of the Latin word *est* (cf. Cajori, *A History of Mathematical Notations*, § 689, sq..)

[29]Whitehead and Russell, *Principia Mathematica*; Martin-Löf, *Intuitionistic Type Theory*.

[30]E.g., in de Bruijn, 'A survey of the project Automath', passim.

[31]Peirce, 'On the Algebra of Logic', § 2.

[32]One example of such an interpretation is the Aczel interpretation ('The type theoretic interpretation of constructive set theory'), where the extensional and accidental notion of set employed in constructive set theory is interpreted in terms of the intensional and essential notion of set. The aforementioned notion of setoid is another example of such an interpretation.

There are four senses in which this definition is not completely exact. First, it is a *descriptive definition*; i.e., it defines the notion of set by listing some characteristic features of it. Next, it is not completely formal since the universal concept mentioned in the definition may contain matter, i.e., signify real being, and, as Aristotle says: "Mathematical accuracy is not to be demanded in everything, but only in things which do not contain matter."[33] Thirdly, it is not completely exact since it fails to distinguish between canonical and noncanonical elements of a set. Definitions 4 and 6, taken together, clarify this. Finally, it is not completely exact in the sense that there is no objective yard-stick against which to measure a proposed set. This is intrinsic to the notion of set, and not remedied by Definitions 4 and 6. Indeed, without drastic measure, this problem seems to be impossible to overcome.

There are a few words in this descriptive definition which perhaps demand some clarifications.

Universal concept. Recall that "by universal we mean that which by nature appertains to several things".[34] St. Thomas makes the following comment on this passage:

"Now it should be noted that he describes a universal as what is naturally disposed to exist in many, and not as what exists in many; because there are some universals which contain under themselves only one singular thing, for example, sun and moon."[35]

With Frege, I admit that a universal concept may even have empty extension, as his example *moon of Venus* shows.[36] Since classical logic, i.e., Aristotelian and medieval logic, only deals with real being, concepts with empty extension are tacitly excluded from it, they are however indispensable to intuitionistic type theory.

First intention. Ordinarily, i.e., outside logic and philosophy, concepts such as *man*, *number*, etc. are used only in predicate position but, by talking *about* concepts, concepts are made *subjects* of the inquiry. For example, consider the sentence *the concept man is not empty.*[37] The subject of this sentence is the concept man, or simply *man*. Philosophers say that the concept man is turned into a reflex concept when used in this way.[38] Reflex concepts are not first intentions. For a universal concept to be a first intention, its content has to be without reference to a thinking mind. Note that when I call a direct concept a first intention, the direct concept has to be understood as the *act* of directing the mind to the thing. For example, the mind first directs its attention to particular animals, plants, houses, colours, numbers, etc., and this act is the first intending, or directing, act.

---

[33]Aristotle, *Metaph.*, Bk. 2, Ch. 3, § 3.

[34]Ibid., Bk. 7, Ch. 13, § 3.

[35]Aquinas, *In Metaph.* Bk. 7, Les. 13, n. 9: "Sciendum autem quod ideo dicit quod universale est quod natum est pluribus inesse, non autem quod pluribus inest; quia quaedam universalia sunt quae non continent sub se nisi unum singulare, sicut sol et luna" (trans. Rowan).

[36]Frege, *Die Grundlagen der Arithmetik*, § 46. Cf. also the distinction between nonsense and absurdity on p. 21 of this book. That something is an element of an empty set is absurd, but it is not nonsense.

[37]Frege, 'On Concept and object', p. 47.

[38]Cf. p. 16.

We think using first intentions when we think without reference to our own thinking. This rules out the set of all concepts, the set of all meaningful expressions,[39] the set of all sets,[40] and similar paradoxical sets.

Consequently, the question whether any universal concept can be made into a set is answered in the negative. The motivation for the restriction to first intentions is similar to the motivation for the ramified theory of types even if there are no third or fourth intentions:

"Yet all these (concepts) are described as second intentions regardless of the fact that one is founded upon another; none of them is ever described as third or fourth intention, because they all belong to the object as known; now, the state of being known is always, for anything, a second state."[41]

The unfortunate axiom of reducibility, or the impredicative axiom, which was introduced into ramified type theory on purely practical grounds, is certainly not justified under the standard interpretation of types "and there is no reason whatever to suppose it true".[42] This axiom states that every higher order type can be reduced to a first order type with the same extension.[43] Set theory has since taken the path of discarding ramification but retaining impredicativity while the approach taken in intuitionistic type theory consists in discarding impredicativity but retaining ramification, at least in the form of the distinction between first and second intentions.

For example, in intuitionistic type theory, Frege's impredicative definition of natural number (discrete quantity) and Dedekind's impredicative definition of real number (magnitude or continuous quantity), under which natural and read numbers are second intentions, have to be discarded in favor of predicative definitions of the same concepts, e.g., some versions of Peano's definition of natural number and Bishop's definition of real number.[44]

---

[39]That *meaningful expression* is a second intention is shown by Berry's paradox (cf. Russell, 'Mathematical Logic as Based on the Theory of Types', § 1). Consider the following question: Does there exist a *least number, that cannot be defined by an expression of at most fifteen words*? On the one had, *yes*, because the number of meaningful expressions of at most fifteen words is finite, and on the other hand, *no*, because if such a number did exist, it would be definable by the words italicized in the question, which is absurd. This version of the paradox is due to Brouwer, 'Intuitionism and formalism', who confuses it with Richard's paradox ('Les Principes des Mathématiques et le Problème des Ensembles').

[40]Cf. p. 57 of this chapter.

[41]Poinsot, *Material Logic*, pp. 73-74. On the other hand, Whitehead and Russell formalize not only first and second order types, but also types of order three, four, etc. *ad infinitum* (*Principia Mathematica*, Intro., Ch. 2). The notion of a second intention should be compared to Frege's notion *second-level concept* (*Begriff zweiter Stufe* in German), 'On Concept and object', p. 49, sq.. Cf. also Martin-Löf, *Intuitionistic Type Theory*, p. 22. The term second intention stems from Avicenna's commentary on Aristotle and is derived from Porphyry's distinction between first and second imposition (Kneale and Kneale, *The Development of Logic*, p. 229, sq.); second intentions were also called logical intentions, since they are studied by logic, and intention became identified with notion by Pacius. Consequently, the three terms second intention, logical intention, and logical notion are interchangeable.

[42]Ramsey, 'Mathematical Logic', p. 186.

[43]Whitehead and Russell, *Principia Mathematica*, n. 12.1, pp. 173–175. The pragmatic justification of the axiom of reducibility is given on p. 62.

[44]These definitions are found in the following places: Frege, *Grundgesetze der Arithmetik I*, § 42; Dedekind, 'Continuity and Irrational Numbers', § 3; Peano, *Arithmetices Principia Nova Methodo Exposita*, § 1; and Bishop and Bridges, *Constructive Analysis*, p. 18. Cf.

Collected to a whole and substantiated. If C is a universal concept, I will abuse Cantor's notation and write $\{C\}$ for the set of individuals falling under C.[45] Moreover, if M is a set, I use el(M) as an abbreviation for element of M. The universal concept C is always equal to the universal concept el($\{C\}$).[46] Moreover, it is $\{C\}$ which is called a set, and not the universal concept C itself. Conversely, if M is a set, then M is not universal, and is spoken of as an object, i.e., as substantiated or hypostasized. Consequently, we cannot say that something is an M, but only that something is an element of M.

For example, the concept *number* is used as a direct concept when we say that 3 is a number, and as a reflex concept when we say that {number} is a set.[47]

Considered as individuals. Here I use the word *individual* in the sense now current in logic.[48] This use of the word is defensible since the elements of a set are considered as atomic and indivisible from the point of view of intuitionistic type theory, even though they may be endowed with additional meaning outside of it. For example, numbers can be used as predicates in natural language, as in *the number of participants were five*, but are considered as individuals from the point of view of intuitionistic type theory.

Definitional equality. Definitional equality is characterized by the four requirements given in the definition, though they do not form a complete characterization.[49] Definitional equality has to be distinguished from propositional equality. That two elements of a set are definitionally equal is a form of assertion, whereas that they are propositionally equal is a proposition.[50]

Definitum—definiens. First a note on terminology, in a definition, which has the general form

$$\underbrace{\text{definitum} \stackrel{\text{def}}{=} \text{definiens}}_{\text{definitio}},$$

the left-hand side is called *definitum* and the right-hand side is called *definiens*.[51] The first requirement for definitional equality is that the two terms, the

Weyl, 'Der circulus vitiosus in der heutigen Begründung der Analysis'; and id., 'Über die neue Grundlagenkrise der Mathematik'.

[45]Cantor introduces this notation, with a slightly different meaning, in 'Beiträge zur Begründung der transfiniten Mengenlehre', p. 481.

[46]Frege make the same point when he says that "someone falling under the concept man" means the same as "a man" ('On Concept and object', p. 47). By saying that the concepts C and el($\{C\}$) are equal, I mean that predication of them amounts to the same.

[47]Cf. Poinsot, *Material Logic*, p. 421 and p. 16 of this book.

[48]See, e.g., Carnap, *Introduction to Symbolic Logic and Its Applications*, p. 4.

[49]These requirements are a slight modification of Martin-Löf's requirements 'About models for intuitionistic type theories and the notion of definitional equality', p. 93. In fn. 7 of 'Über eine bisher noch nicht benütze Erweiterung des finiten Standpunktes', Gödel remarks that identity between functions is to be understood as intensional or definitional equality, this seems to be the origin of the term *definitional.* Cf. de Bruijn, *Automath, a language for mathematics*, p. 28.

[50]Cf. Martin-Löf, *Intuitionistic Type Theory*, p. 31. (However, a distinction is to be made between synonymy, i.e., identity of meaning, and definitional equality.) Cf. p. 33 of this book.

[51]The terminology favored by the authors of *Principia Mathematica*, Intro., Ch. 1, p. 11, is *definiendum* and *definiens.* I prefer the classical terminology since the left-hand side is not *supposed to be defined* by the definition, rather it *is defined* by it. For the whole equality, I have used the Latin word *definitio*, since the word *definition* is used confusedly between

definitum and the definiens, of a definition really are definitionally equal. This is of course why this equality relation is called definitional equality.

Form—parts. The second requirement states that if two elements have the same form, e.g., $f(p_1, \ldots, p_n)$ and $f(q_1, \ldots, q_n)$, and their parts are equal, i.e., p_1 and q_1 are equal objects of the logical category demanded by the form f, etc. then $f(p_1, \ldots, p_n)$ and $f(q_1, \ldots, q_n)$ are equal elements of the set in question.

Any object is equal to itself. The third requirement is that equality be reflexive. A lot has been written concerning reflexivity in the form of the principle of identity, starting with Plato's dialog Parmenides. Instead of entering into metaphysical considerations I here take the approach of considering a universal concept for which equality is not reflexive unfit as a basis for a set and leave open the question whether or not there are such universal concepts.

Two objects which equal a third etc. The fourth requirement is a variant of Euclid's first common notion: "things which equal the same thing also equal one another".[52] An equality relation satisfying this condition will be called *cancellable*, i.e., definitional equality between elements of a set has to be cancellable.

This completes the clarifications of Definition 3.

§ 4. Canonical sets and elements

The definition given in the preceding section is a *descriptive definition*. By this I mean that it tries to capture the notion of set by listing various characteristic features of it. Some people will say that it lacks exactness, others will be perfectly happy with it: "Again, some require exactness in everything, while others are annoyed by it, either because they cannot follow the reasoning or because of its pettiness; for there is something about exactness which seems to some people to be mean, no less in an argument than in a business transaction."[53] In developing an exact and formal language, a *lingua characteristica*, one inevitably encounters great difficulties at the very outset: "one cannot proceed from the informal to the formal by formal means".[54] The most basic, or primitive, notions cannot be defined with the exactness and rigor expected from nominal definitions inside the formal language itself, since this would lead to an infinite regress. To explain, or to define, these primitive notions is to boot-strap the formal language, or to perform Baron Münchhausen's trick.

Descriptive definitions sometimes capture the concept defined exactly, and sometimes not; they can even be inconsistent by including features which turn out to be contradictory. It is clear that a third kind of definition, between the nominal definition, which is exact but adds no content, and the descriptive definition, which is inexact and runs the risk of adding content which should not be there, is needed. This is where the type-theoretic *meaning explanations* come in; a meaning explanation always defines precisely what a form of assertion *means*. That is, it does not attempt to capture a pre-existing concept by

---

definiens and *definitio* in ordinary speech.

[52]Euclid, *Elementa*, Bk. 1, cf. also Aristotle, *Top.*, Bk. 7, Ch. 1, 152a30.

[53]Aristotle, *Metaph.*, Bk. 2, Ch. 3, § 2.

[54]Perlis, 'Epigrams on Programming', n. 102.

describing its content, but introduces a new concept by an exact explanation of its content.

To use yet another figure of speech, in Martin-Löf's original formulations of intuitionistic type theory, the Gordian knot is cut by the explanation of the notion of set.[55] In a like manner, in the present formulation of intuitionistic type theory, the Gordian know is cut by the following definition of the notion of set, which, together with Definition 6 on p. 91, forms an elaboration of the previous descriptive definition.

Definition 4. *That A is a set, abbreviated A ε* set, *means four things: that it is defined when a is an element of the set A, abbreviated a ε* el(A), *in any way whatever, but always without reference to the totality of sets; that it is defined when two elements a and b of the set A are definitionally equal, abbreviated $a = b$ ε* el(A)*; that an element of the set in question is always equal to itself; and that two elements which equal a third element of the set in question are equal to one another.*

This defines the form of assertion A ε set and, as it were, as a side effect, the forms of assertion a ε el(A) and $a = b$ ε el(A) are defined as well.[56] That is, their meanings are postulated to be dependent on the definition of A and found in *its* definition. That a is an element of A presupposes that A is a set and its meaning is determined by A. Similarly, that a and b are equal elements of A presupposes that a and b are elements of A which in turn presuppose that A is a set, and its meaning is also determined by the set A.

Only *canonical* sets and elements are defined by this definition. I say canonical to distinguish a term which immediately refers to a set or element from a term which refers only through computation, which, in this context, is called a *noncanonical* set or element. This distinction applies to expressions and to their meanings, but not to the objects denoted by the expressions.[57] For example, 4 is a canonical decimal number, but $2+2$ is a noncanonical decimal number. In the text, it will in most cases be clear from the context if set and element mean canonical or noncanonical set and element. In the formal language, however, the epsilon sign (ε) will be used as copula in assertions involving canonical sets and elements, whereas the colon sign (:) will be used as copula in assertions involving noncanonical sets and elements.

Both Definition 3 and the above definition are predicative in the sense of Russell and Poincaré.[58] In Definition 3, predicativity is expressed by the phrase *first intention* and in the above definition by the phrase *without reference to the totality of sets.* A predicative definition is noncircular, or well-founded; the notion defined is not in any way presupposed in its definiens. One could say that

---

[55]Martin-Löf, *Intuitionistic Type Theory*, p. 8, quoted on p. 55 of this book.

[56]With respect to the analysis of an expression into form and parts (p. 8), A ε set is analysed as having the sole part A; a ε el(A) is analysed as having two parts, a and A; and $a = b$ ε el(A) is analysed as having three parts, a, b, and A.

[57]Cf. p. 21.

[58]The word predicative was introduced by Russell, 'On some difficulties in the theory of transfinite numbers and order types', p. 34, and commented upon by Poincaré, 'Les mathématiques et la logique', §§ 7–10. For the connection with Poincaré, see further McLarty, 'Poincaré: Mathematics & Logic & Intuition', p. 106. The danger of impredicative definitions in set theory is emphasized by Skolem, 'Some remarks on axiomatized set theory', p. 297.

impredicativity in definition corresponds to a vicious circle, or *petitio principii*, in demonstration.[59] The condition of predicativity is necessarily somewhat vague: some set forming operations are such that one has to make up one's own mind whether to accept them or not.[60]

Now some comments on the latter two parts of this definition. That an element of the set in question is always equal to itself means that if $a \mathrel{\varepsilon} \mathrm{el}(A)$ then $a = a \mathrel{\varepsilon} \mathrm{el}(A)$, i.e., the inference rule

$$\frac{a \mathrel{\varepsilon} \mathrm{el}(A)}{a = a \mathrel{\varepsilon} \mathrm{el}(A)} \tag{D3.1}$$

is a part of $A \mathrel{\varepsilon}$ set means. That two elements which equal a third element of the set in question are equal to one another means that if $a = c \mathrel{\varepsilon} \mathrm{el}(A)$ and $b = c \mathrel{\varepsilon} \mathrm{el}(A)$, then $a = b \mathrel{\varepsilon} \mathrm{el}(A)$, i.e., the inference rule

$$\frac{a = c \mathrel{\varepsilon} \mathrm{el}(A) \quad b = c \mathrel{\varepsilon} \mathrm{el}(A)}{a = b \mathrel{\varepsilon} \mathrm{el}(A)} \tag{D3.2}$$

is also a part of what $A \mathrel{\varepsilon}$ set means. When using the latter inference rule, I will call c the middle term of the inference, by analogy with syllogistic middle terms.

The first two parts of the definition do not give rise to any inference rules. For example, there is no inference rule like

$$\frac{A \mathrel{\varepsilon} \text{set}}{\text{it is defined what } a \mathrel{\varepsilon} \mathrm{el}(A) \text{ means}},$$

simply because the conclusion is not of a form subject to type-theoretical treatment. Still, granted that A is a set, it must be defined what the forms of assertion $a \mathrel{\varepsilon} \mathrm{el}(A)$ and $a = b \mathrel{\varepsilon} \mathrm{el}(A)$ mean, for this particular set A.

Inference rules which are evident from the meanings of the terms involved without need of further explanations are called *meaning determining*. For example, knowing that a is an element of A, we know in particular the presupposition, that A is a set, and this means, by the third part of the definition, that a is equal to itself; that is, inference rule (D3.1) is (partly) meaning determining for $A \mathrel{\varepsilon}$ set. This becomes even clearer if the presupposition is spelled out as a first premiss:

$$\frac{A \mathrel{\varepsilon} \text{set} \quad a \mathrel{\varepsilon} \mathrm{el}(A)}{a = a \mathrel{\varepsilon} \mathrm{el}(A)}.$$

In the same way, (D3.2) is meaning determining for $A \mathrel{\varepsilon}$ set.

Instead of the usual requirements of reflexivity, symmetry and transitivity, I require that the equality defined on A be reflexive and cancellable, i.e., that two elements which equal a third element are equal to one another.[61] But it can

[59] Cf. Whitehead and Russell, *Principia Mathematica*, Intro., Ch. 2, § 1, p. 40; and Aristotle, *An. Pr.*, Bk. 1, Ch. 23.

[60] One famous example of such a set forming operation is the Mahlo-universe contrived by Setzer, 'Extending Martin-Löf's Type Theory by one Mahlo-universe'.

[61] Reflexivity, symmetry, and transitivity are the standard requirements (Martin-Löf, *Intuitionistic Type Theory*, p. 8, and p. 14). To be exact, one should say cancellable from the right, but, as the reader may verify, if a relation is reflexive and cancellable from one side, it is also cancellable from the other side.

be shown that the relation of equality on a set A is symmetric and transitive. That is, the inference rules

$$\frac{a = b \ \varepsilon \ \mathrm{el}(A)}{b = a \ \varepsilon \ \mathrm{el}(A)}, \tag{M3.1}$$

and

$$\frac{a = b \ \varepsilon \ \mathrm{el}(A) \quad b = c \ \varepsilon \ \mathrm{el}(A)}{a = c \ \varepsilon \ \mathrm{el}(A)}, \tag{M3.2}$$

are both valid. This is demonstrated by the schematic demonstrations

$$\dfrac{\dfrac{b \ \varepsilon \ \mathrm{el}(A)}{b = b \ \varepsilon \ \mathrm{el}(A)}\,(\mathrm{D3.1}) \qquad a = b \ \varepsilon \ \mathrm{el}(A)}{b = a \ \varepsilon \ \mathrm{el}(A)}\,(\mathrm{D3.2}),$$

where the premiss $b \ \varepsilon \ \mathrm{el}(A)$ is just a presupposition of $a = b \ \varepsilon \ \mathrm{el}(A)$, and

$$\dfrac{a = b \ \varepsilon \ \mathrm{el}(A) \quad \dfrac{b = c \ \varepsilon \ \mathrm{el}(A)}{c = b \ \varepsilon \ \mathrm{el}(A)}\,(\mathrm{M3.1})}{a = c \ \varepsilon \ \mathrm{el}(A)}\,(\mathrm{D3.2}).$$

Conversely, if a relation is reflexive, symmetric and transitive, then it is cancellable.

Inference rules like symmetry and transitivity are called mediate inference rules. Immediate and mediate inference rules are valid in equal measure, but for different reasons. Once one gets used to a particular pattern, or schema, of demonstration, one can spell it out as a mediate inference rule. Thus, mediate inference rules are indispensable for convenient demonstration—mentally as well as on paper. One has to remember only the mediate inference rule, not its schematic demonstration. Any use of a mediate inference rule is an abbreviation of, or substitute for, its schematic demonstration.[62]

The next topic is equality between sets. The fourth basic form of assertion, that two sets A and B are equal, presupposes that A and B are sets and its meaning is given by the following definition.

Definition 5. *That A and B are equal sets, abbreviated $A = B \ \varepsilon$ set, means four things: if $a \ \varepsilon \ \mathrm{el}(A)$ then $a \ \varepsilon \ \mathrm{el}(B)$, if $a = b \ \varepsilon \ \mathrm{el}(A)$ then $a = b \ \varepsilon \ \mathrm{el}(B)$; conversely, if $a \ \varepsilon \ \mathrm{el}(B)$ then $a \ \varepsilon \ \mathrm{el}(A)$, and if $a = b \ \varepsilon \ \mathrm{el}(B)$ then $a = b \ \varepsilon \ \mathrm{el}(A)$.*

The above definition, which is somewhat repetitive, can be summarized by

$$A = B \ \varepsilon \ \text{set} \quad \text{means that} \quad \frac{a \ \varepsilon \ \mathrm{el}(A)}{a \ \varepsilon \ \mathrm{el}(B)} \quad \text{and} \quad \frac{a = b \ \varepsilon \ \mathrm{el}(A)}{a = b \ \varepsilon \ \mathrm{el}(B)},$$

where the double line indicates that the inference in question is valid in both directions.[63] Spelling this out, the following four inference rules are immediate

[62]Cf. Husserl, *Log. Unt. I*, § 6: "I know that the Pythagorean theorem is true—I can prove it" vs. "—but I have forgotten the proof" (trans. Findlay). Cf. ibid., § 9; and Rule 7 of Descartes' 'Rules for the Direction of the Mind'. I will use the phrase *schematic demonstration* for that which shows the validity of a mediate inference rule.

[63]Cf. Martin-Löf, *Intuitionistic Type Theory*, p. 9.

from the definition of equality between sets:

$$\frac{A = B \ \varepsilon \ \text{set} \quad a \ \varepsilon \ \text{el}(A)}{a \ \varepsilon \ \text{el}(B)}, \tag{D3.3}$$

and

$$\frac{A = B \ \varepsilon \ \text{set} \quad a = b \ \varepsilon \ \text{el}(A)}{a = b \ \varepsilon \ \text{el}(B)}, \tag{D3.4}$$

and conversely,

$$\frac{A = B \ \varepsilon \ \text{set} \quad a \ \varepsilon \ \text{el}(B)}{a \ \varepsilon \ \text{el}(A)}, \tag{D3.5}$$

and

$$\frac{A = B \ \varepsilon \ \text{set} \quad a = b \ \varepsilon \ \text{el}(B)}{a = b \ \varepsilon \ \text{el}(A)}. \tag{D3.6}$$

These rules of inference are meaning determining for the form of assertion $A = B \ \varepsilon$ set. I will call them rules of set conversion.

It also follows from the definition that equality between sets is cancellable and reflexive. That is, the inference rules

$$\frac{A = C \ \varepsilon \ \text{set} \quad B = C \ \varepsilon \ \text{set}}{A = B \ \varepsilon \ \text{set}} \tag{J3.1}$$

and

$$\frac{A \ \varepsilon \ \text{set}}{A = A \ \varepsilon \ \text{set}} \tag{J3.2}$$

are both valid.

Justification of (J3.1). First, let the premisses $A = C \ \varepsilon$ set and $B = C \ \varepsilon$ set be given. We want to know that $A = B \ \varepsilon$ set. There are four things to establish, corresponding to the four parts of the definition of $A = B \ \varepsilon$ set. First, we want to know that if $a \ \varepsilon \ \text{el}(A)$ then $a \ \varepsilon \ \text{el}(B)$. Thus, let $a \ \varepsilon \ \text{el}(A)$ be given. Reason as follows: $A = C \ \varepsilon$ set and $a \ \varepsilon \ \text{el}(A)$, therefore $a \ \varepsilon \ \text{el}(C)$, by (D3.3); $B = C \ \varepsilon$ set and $a \ \varepsilon \ \text{el}(C)$, therefore $a \ \varepsilon \ \text{el}(B)$, by (D3.5), as required. Next, we want to know that if $a = b \ \varepsilon \ \text{el}(A)$ then $a = b \ \varepsilon \ \text{el}(B)$. Reason as follows: $A = C \ \varepsilon$ set and $a = b \ \varepsilon \ \text{el}(A)$, therefore $a = b \ \varepsilon \ \text{el}(C)$, by (D3.4); $B = C \ \varepsilon$ set and $a = b \ \varepsilon \ \text{el}(C)$, therefore $a = b \ \varepsilon \ \text{el}(B)$, by (D3.6), as required. The other two parts of the justification are similar, but instead use the inference rules in the other order. This completes the justification.

Justification of (J3.2). Having understood the way in which an inference rule is justified, the validity of this inference rule is trivial.

Inference rules such as (J3.2) and (J3.1) will be called *justified inference rules*. The difference between a meaning determining inference rule, a justified inference rule, and a mediate inference rule is the following. While a meaning determining inference rule is evident from the meaning of the terms involved without need of further explanation, justified inference rules need some explanation. The difference between meaning determining and justified inference rules on the one hand, and mediate inference rules on the other hand, is that while the latter are shown to be valid by a schematic demonstration *formulated in the language* of intuitionistic type theory, the former are self-evident. The purpose of the justification is to expound the meaning of the terms involved as to make the assertion or inference rule intuitively evident. We speak of *intuitive* validity both for meaning determining and justified inference rules. Here intuition is not to be understood as a vague feeling, but rather as a certain intellectual perception.[64]

Having justified the above two inference rules, symmetry and transitivity of set equality become mediate inference rules:

$$\frac{A = B \ \varepsilon \ \text{set}}{B = A \ \varepsilon \ \text{set}}\,, \tag{M3.3}$$

and

$$\frac{A = B \ \varepsilon \ \text{set} \quad B = C \ \varepsilon \ \text{set}}{A = C \ \varepsilon \ \text{set}}\,. \tag{M3.4}$$

The schematic demonstrations are the same, *mutatis mutandis*, as those for equality between elements. However, one could instead justify symmetry and transitivity directly, and let cancellability be a mediate inference rule;[65] this makes little difference, but some choice has to be made. A guiding principle can be Ockham's law of parsimony: *entia non sunt multiplicanda praeter necessitatem*, according to which the number of justified inference rules should be minimized.

Comparing the mediate inference rule (M3.3) with the meaning determining inference rules (D3.3)–(D3.6), it seems as if inference rules (D3.5) and (D3.6) are redundant, since, e.g., (D3.5) could be demonstrated by

$$\frac{\dfrac{A = B \ \varepsilon \ \text{set}}{B = A \ \varepsilon \ \text{set}}\,(\text{M3.3}) \qquad a \ \varepsilon \ \text{el}(B)}{a \ \varepsilon \ \text{el}(A)}\,(\text{D3.3}).$$

This is however not the case. Rather, inference rule (M3.3) is valid *because* the definition of equality between sets $A = B \ \varepsilon$ set is symmetric in A and B. The four inference rules (D3.3)–(D3.6) are valid *before* (M3.3) is valid, and the validity of the meaning determining inference rules is the cause of the validity of the justified inference rules from which the mediate inference rule is demonstrated.

The notion of equality between sets is naturally divided into two parts. Let us say that a set A is a *subset* of the set B, and write $A \sqsubseteq B$, if any element of A is an element of B and any two equal elements of A are equal elements of B. According to this definition, equality between sets is tantamount to mutual inclusion. That is, the inference rule

$$\frac{A \sqsubseteq B \quad B \sqsubseteq A}{A = B \ \varepsilon \ \text{set}}$$

is valid, as are the inference rules

$$\frac{A = B \ \varepsilon \ \text{set}}{A \sqsubseteq B} \quad \text{and} \quad \frac{A = B \ \varepsilon \ \text{set}}{B \sqsubseteq A}\,.$$

---

[64] I use the word in the same sense as Descartes, 'Rules for the Direction of the Mind', Rule 3, q.v.

[65] This phenomenon can be related to Geach's insight that, just because it is necessary to take some propositions as undemonstrated, it does not follow that there are some propositions which it is necessary to take as undemonstrated (*Logic Matters*, pp. 4–5, originally published as Geach, 'History of a Fallacy').

I have chosen to take equality instead of inclusion as the basic notion since that is how it is usually done.

§ 5. How to define a canonical set

An assertion of the form $A \varepsilon$ set, for a particular form of set A, is recognized as valid after the four things required in the definition of $A \varepsilon$ set have been done. Consider the set of Booleans, which I will write B. For the elements of the set B, I will adopt Boole's notation 1 for true and 0 for false.[66] First I claim that

$$\mathrm{B}\ \varepsilon\ \text{set}. \tag{R3.1}$$

Next I have to perform the four steps required by the above definition of the notion of set. That m is an element of the set B means that m is either 1 or 0, i.e.,

$$1\ \varepsilon\ \mathrm{el(B)}, \tag{D3.7}$$

and

$$0\ \varepsilon\ \mathrm{el(B)}; \tag{D3.8}$$

that m and n are equal elements of the set B means that they are either both 1 or both 0, i.e., the assertions $1 = 1\ \varepsilon\ \mathrm{el(B)}$ and $0 = 0\ \varepsilon\ \mathrm{el(B)}$ are both meaning determining;[67] this equality relation is evidently reflexive and cancellable.

Now, what is the status of the assertion B ε set ? It is an assertion which is recognized as valid because something *has been done.* The situation is comparable to the mathematical practice of writing down the theorem before the proof, while in fact it is the proof that makes the alleged theorem into a theorem. In our case, the situation is even more complicated since $1\ \varepsilon\ \mathrm{el(B)}$ presupposes that B ε set and $1 = 1\ \varepsilon\ \mathrm{el(B)}$ presupposes that $1\ \varepsilon\ \mathrm{el(B)}$. Thus, all steps in the definition of a set have to be understood together. The general pattern is as follows:

Introduce the new form of set to be defined.

Define what it means to be an element of the set in question.

Define what it means for two elements of the set to be equal.

Make sure that the equality relation so defined is reflexive.

Make sure that the equality relation so defined is cancellable.

I have adopted Martin-Löf's terminology for the assertions and inference rules due to the first, second, and third steps; they are called, respectively, rules of set formation, introduction rules, and equality rules.[68] The introduction and equality rules are always meaning determining. But what about the rule of set formation ? It is a rule which is *recognized* as valid because something *has been done.* Here we encounter a forth kind of validity. The complete list is now given by: meaning determining, mediate, justified, and recognized, cf. Table 5. To

[66]Strictly speaking, the Booleans of Boole, written 1 and 0, stand for Universe and Nothing (*An Investigation of the Laws of Thought*, p. 34). An alternative notation, due to Peirce ('On the Algebra of Logic: a contribution to the philosophy of notation', § 2) is to write v for *verum* and f for *falsum*; in English translation, this becomes t and f, as in the programming language Scheme. In addition to the interpretation of 1 and 0 as truth values, they can also be interpreted as bits in a digital computer, in which case 1 is identified with the on-bit and

- A valid assertion or inference rule is either
  - (M) *mediate* by exhibiting a schematic demonstration; or
  - an *axiom*, i.e., *self-evident* or *immediate*; a self-evident assertion or inference rule is either:
    - (D) a meaning determining axiom or inference rule, i.e., an axiom or inference rule which is self-evident without further need of explanation; or
    - an assertion or inference rule which is self-evident after the meanings of the terms have been unfolded, of two kinds:
      - (J) a *justified* assertion or inference rule, where the meanings of the terms involved are expounded as to make the assertion or inference rule intuitively evident; or
      - (R) a *recognized* assertion or inference rule, for which the relevant meaning intentions of the assertion or inference rule are fulfilled.

Table 5. Classification of valid assertions and inference rules, according to the reason for their validity.

assertions and inference rules which are *recognized* as valid, it seems suitable to apply Husserlian terminology and say that their meaning intentions have been fulfilled.[69] That is, we *recognize* that B is a canonical set, not because the meanings of the terms are expounded in a justification, but simply because we recognize that the meaning intentions laid down in definition 4 have been fulfilled.

Something should also be said about the last two steps in recognizing the validity of a rule of set formation, viz., the making sure that the equality relation so defined is reflexive and cancellable. This *making sure* is not to be understood as a justification of inference rules (D3.1) and (D3.2) for the set in question. Rather, in one way or another, this *making sure* should be founded on an intrinsic connection between the meaning of $a \, \varepsilon \, \mathrm{el}(A)$ and the meaning of $a = a \, \varepsilon \, \mathrm{el}(A)$ for reflexivity, and between the meanings of $a = c \, \varepsilon \, \mathrm{el}(A)$ and $b = c \, \varepsilon \, \mathrm{el}(A)$ together, and the meaning of $a = b \, \varepsilon \, \mathrm{el}(A)$ for cancellability. It is in recognizing that A is a set that reflexivity and cancellability are promoted to inference rules valid *simpliciter*, because they then form a part of what it means for A to be a set.

A mathematical set is often defined by enumerating the forms that the elements may have. That is, a set A is often defined by specifying a number of forms $f_1, \ldots, f_n$ which the elements may have, called constructors, each with a certain arity. Moreover, two elements are typically considered equal if they have the same form and their respective parts are equal objects of the relevant logical

---

0 with the off-bit.

[67]These two axioms are not given any numbers since, when used, they can be taken as instances of the general reflexivity rule (D3.1).

[68]Martin-Löf, *Intuitionistic Type Theory*, p. 24.

[69]Cf. Husserl, *Log. Unt. II*, Pt. 2, Inv. 6, Ch. 1.

category. In mathematical logic, such a set is often called *inductive*. In general, the word *induction* signifies the process of inferring a general principle from particular instances. A special case of induction is mathematical, or complete, induction, which is a logically sound mood of demonstration. Accordingly, a set is called inductive if it admits proof by complete induction. The set of Booleans is inductive in this sense, and I now proceed to define some of the most important inductive sets.

The natural numbers are central to arithmetic, the Queen of mathematics, whence it is important that they form a set.[70] We must therefore perform the four steps required to fulfill the meaning intentions of the assertion

$$\mathrm{N}\ \varepsilon\ \mathrm{set}. \tag{R3.2}$$

First we have to define what it means for something to be an element of the set N. An element of N is either 0 or s(n), where n already is an element of N, i.e., we have the introduction rules

$$0\ \varepsilon\ \mathrm{el}(\mathrm{N}), \tag{D3.9}$$

and

$$\frac{n\ \varepsilon\ \mathrm{el}(\mathrm{N})}{\mathrm{s}(n)\ \varepsilon\ \mathrm{el}(\mathrm{N})}\ . \tag{D3.10}$$

As required by an inductive set, equality between elements consists in equality between form and parts, i.e., we have the equality rules $0 = 0\ \varepsilon\ \mathrm{el}(\mathrm{N})$, and

$$\frac{n = m\ \varepsilon\ \mathrm{el}(\mathrm{N})}{\mathrm{s}(n) = \mathrm{s}(m)\ \varepsilon\ \mathrm{el}(\mathrm{N})}\ . \tag{D3.11}$$

To complete the required four steps, we must make sure that the equality relation so defined is reflexive and cancellable, but this is trivial since the equality rules, as it were, mirror the introduction rules.

Another basic set of some importance is the set with only one element. This set is called the *unit* set, and will be denoted by 1.

$$1\ \varepsilon\ \mathrm{set}. \tag{R3.3}$$

The sole element of 1 will be written 0, i.e.,[71]

$$0\ \varepsilon\ \mathrm{el}(1). \tag{D3.12}$$

Since equality between elements is trivial, i.e., defined by $0 = 0\ \varepsilon\ \mathrm{el}(1)$, we recognize assertion (R3.3) as valid.

One of the more intriguing sets in intuitionistic type theory is the empty set, denoted $\emptyset$.[72] We have

$$\emptyset\ \varepsilon\ \mathrm{set}. \tag{R3.4}$$

The definition of $a\ \varepsilon\ \mathrm{el}(\emptyset)$ consists in saying that there are no elements of the empty set. This could be formulated in several different ways. One way to put it is that $\emptyset$ is an inductive set with *zero* forms of elements; another formulation is that $\emptyset$ is the set that has no elements. In any case, equality between elements

---

[70]This definition of the natural numbers is essentially due to Peano, *Arithmetices Principia Nova Methodo Exposita*, with the difference that we start at zero instead of at one.

[71]An alternative notation, popular in category theory, is $\star$, i.e., $\star\ \varepsilon\ \mathrm{el}(1)$, and $1 = \{\star\}$.

[72]The use of the symbol $\emptyset$, taken from the Norwegian alphabet, for the empty set is due to Weil (*The Apprenticeship of a Mathematician*, p. 144).

of the empty set is trivial, since there are no elements, and (R3.4) is recognized as valid.

The Cartesian, or direct, product $A \times B$ of two sets A and B is our first example of a set-forming operation. The product is called Cartesian, of course, after Descartes, because it is a generalization of Descartes' discovery that the Euclidean plane can be viewed as a kind of product of two lines.[73] The generalization lies in allowing the kinds of objects put on the axes, here elements of A and B, respectively, to vary. In the general case, the co-ordinates of a point in the plane is simply a pair of an element of A and an element of B. If we take both A and B to be the set of real numbers, we get back the plane of Descartes' analytic geometry. If both A and B are finite sets, the number of elements in the set $A \times B$ is equal to the number of elements in A times the number of elements in B, so the use of the product sign for the Cartesian product is suggestive. The formation rule for the Cartesian product is given by

$$\frac{A \;\varepsilon\; \text{set} \quad B \;\varepsilon\; \text{set}}{A \times B \;\varepsilon\; \text{set}} \,. \tag{R3.5}$$

The form '$\times$' of the conclusion is called a set former. Granted that A and B are sets, we have to do the usual four things. First, an element of $A \times B$ has the general form $(a\,,b)$, where $a \;\varepsilon\; \text{el}(A)$ and $b \;\varepsilon\; \text{el}(B)$, i.e., we have the introduction rule

$$\frac{a \;\varepsilon\; \text{el}(A) \quad b \;\varepsilon\; \text{el}(B)}{(a\,,b) \;\varepsilon\; \text{el}(A \times B)} \,, \tag{D3.13}$$

and this is the only form which a canonical element of $A \times B$ can have. Next, two elements $(a\,,b)$ and $(c\,,d)$ are equal if a and c are equal elements of A and b and d are equal elements of B, i.e., we have the equality rule

$$\frac{a = c \;\varepsilon\; \text{el}(A) \quad b = d \;\varepsilon\; \text{el}(B)}{(a\,,b) = (c\,,d) \;\varepsilon\; \text{el}(A \times B)} \,. \tag{D3.14}$$

Finally, this equality relation is clearly reflexive and cancellable, since the equality relations on the sets A and B are reflexive and cancellable. This completes the four steps. In addition, we have the inference rule

$$\frac{A = C \;\varepsilon\; \text{set} \quad B = D \;\varepsilon\; \text{set}}{A \times B = C \times D \;\varepsilon\; \text{set}} \,. \tag{J3.3}$$

It is typical for set-forming operations, such as the Cartesian product, that one has to verify that the set former in fact respects equality between sets, so that the general principle that two complex terms are equal if their parts are equal is maintained. In this case, the justification is left to the reader as a straightforward but tedious exercise.

The sum of two sets A and B, also called the disjoint union, or coproduct, is denoted $A + B$.[74] As suggested by both terminology and notation, if both A and B are finite sets, then the number of elements in the set $A + B$ is equal to

[73]Cf. Descartes, *La géométrie*; this work was published in 1637 as an appendix to his *Discours de la Méthode.*

[74]Martin-Löf, *Intuitionistic Type Theory*, p. 55. The notations $\text{i}(a)$ and $\text{j}(b)$ are from the same book, but there are several notational variants in the literature. These letters are chosen because i and j are *injections.*

the number of elements in A plus the number of elements in B. We have the formation rule

$$\frac{A\ \varepsilon\ \mathrm{set} \quad B\ \varepsilon\ \mathrm{set}}{A+B\ \varepsilon\ \mathrm{set}}\ . \tag{R3.6}$$

An element of $A+B$ either has the form $\mathrm{i}(a)$, where $a\ \varepsilon\ \mathrm{el}(A)$, or the form $\mathrm{j}(b)$, where $b\ \varepsilon\ \mathrm{el}(B)$, i.e., we have the introduction rules

$$\frac{a\ \varepsilon\ \mathrm{el}(A) \quad (B\ \varepsilon\ \mathrm{set})}{\mathrm{i}(a)\ \varepsilon\ \mathrm{el}(A+B)}\ , \tag{D3.15}$$

and

$$\frac{(A\ \varepsilon\ \mathrm{set}) \quad b\ \varepsilon\ \mathrm{el}(B)}{\mathrm{j}(b)\ \varepsilon\ \mathrm{el}(A+B)}\ . \tag{D3.16}$$

I put a premiss within parentheses if it is needed only as a presupposition of the conclusion, i.e., to make the inference rule well-formed. As for any inductive set, two elements are equal if they have equal form and equal parts, i.e., we have the equality rules

$$\frac{a=b\ \varepsilon\ \mathrm{el}(A) \quad (B\ \varepsilon\ \mathrm{set})}{\mathrm{i}(a)=\mathrm{i}(b)\ \varepsilon\ \mathrm{el}(A+B)}\ , \tag{D3.17}$$

and

$$\frac{(A\ \varepsilon\ \mathrm{set}) \quad a=b\ \varepsilon\ \mathrm{el}(B)}{\mathrm{j}(a)=\mathrm{j}(b)\ \varepsilon\ \mathrm{el}(A+B)}\ . \tag{D3.18}$$

It is also easy to justify the inference rule

$$\frac{A=C\ \varepsilon\ \mathrm{set} \quad B=D\ \varepsilon\ \mathrm{set}}{A+B=C+D\ \varepsilon\ \mathrm{set}}\ . \tag{J3.4}$$

The next set I will consider is a new formulation of *lists*. The standard list in intuitionistic type theory is the cons-list, or singly-linked list, ubiquitous is functional programming. My suggestion is to instead consider lists of a certain length n. The formation rule for the set of lists of length n is given by

$$\frac{A\ \varepsilon\ \mathrm{set} \quad n\ \varepsilon\ \mathrm{el}(\mathrm{N})}{\mathrm{L}(A,n)\ \varepsilon\ \mathrm{set}}\ . \tag{R3.7}$$

The meaning of the form of assertion $l\ \varepsilon\ \mathrm{el}(\mathrm{L}(A,n))$ depends on the form of n: if n is zero the only possible form of l is nil, which I will write as an empty pair of parentheses; if n has the form $\mathrm{s}(p)$, an element of $\mathrm{L}(A,n)$ is a pair of an element a of A and a list l of length p, which I will write $(a\,,l)$. That is, the inference rules

$$\frac{(A\ \varepsilon\ \mathrm{set})}{()\ \varepsilon\ \mathrm{el}(\mathrm{L}(A,0))} \tag{D3.19}$$

and

$$\frac{a\ \varepsilon\ \mathrm{el}(A) \quad l\ \varepsilon\ \mathrm{el}(\mathrm{L}(A,p))}{(a\,,l)\ \varepsilon\ \mathrm{el}(\mathrm{L}(A,\mathrm{s}(p)))} \tag{D3.20}$$

characterize what it means to be an element of $\mathrm{L}(A,n)$. Equality between elements of $\mathrm{L}(A,n)$ is equality between form and parts, i.e.,

$$\frac{(A\ \varepsilon\ \mathrm{set})}{()=()\ \varepsilon\ \mathrm{el}(\mathrm{L}(A,0))} \tag{D3.21}$$

and

$$\frac{a = b \ \varepsilon \ \mathrm{el}(A) \quad l = m \ \varepsilon \ \mathrm{el}(\mathrm{L}(A, p))}{(a\,, l) = (b\,, m) \ \varepsilon \ \mathrm{el}(\mathrm{L}(A, \mathrm{s}(p)))} \ . \tag{D3.22}$$

As before, the inference rule

$$\frac{A = B \ \varepsilon \ \mathrm{set} \quad p = q \ \varepsilon \ \mathrm{el}(\mathrm{N})}{\mathrm{L}(A, p) = \mathrm{L}(B, q) \ \varepsilon \ \mathrm{set}} \tag{J3.5}$$

is easy to justify.

This completes the definition of some of the most important sets of intuitionistic type theory; there are several other important sets, but these cannot be defined until after noncanonical sets and elements have been introduced.

§ 6. More canonical sets

As a sort of appendix to the previous section, I will present two sets which do not belong to the standard sets of intuitionistic type theory, but which nevertheless are of some interest. These are the set D of decimal numbers and the set E of numbers in the sense of Euclid.

Even if a Peano style definition of the natural numbers has the advantage of making the principle of mathematical induction immediately evident, it has the distinct disadvantage of forgoing a very important discovery, viz., the positional system. We cannot say that, e.g., 4 is a canonical element of N and that it is only an abbreviation for s(s(s(s(0)))) since abbreviatory definitions introduce noncanonical elements.[75] Thus, if we want to use canonical decimal numbers, we have to define another set of natural numbers with its own introduction and equality rules. Using the letter D for the set of decimal numbers, I have to perform the steps required to recognize the assertion

$$\mathrm{D} \ \varepsilon \ \mathrm{set}$$

as valid. It is easiest to exclude zero from the decimal numbers, since then any decimal number can be analysed as having its rightmost digit as its outermost form. For example, the number 410 is analysed as having form 0 and 41 as its sole part; the number 41 has 1 as form and 4 as its sole part; the number 4 has no parts, i.e., it is a *categorem*. If we write c_0 for the unary form corresponding to 0 and c_1 for the unary form corresponding to 1, the 'standardized' notation for the number 410 becomes $\mathrm{c}_0(\mathrm{c}_1(4))$, directly bringing out the outermost form of the expression. The leftmost, or innermost, digit cannot be zero since we have excluded zero. This means that every decimal number can be uniquely written in this form. So the nullary constructors become

$$1 \ \varepsilon \ \mathrm{el}(\mathrm{D}), \qquad \cdots \qquad 9 \ \varepsilon \ \mathrm{el}(\mathrm{D}),$$

and the unary constructors give rise to the inference rules

$$\frac{a \ \varepsilon \ \mathrm{el}(\mathrm{D})}{\mathrm{c}_0(a) \ \varepsilon \ \mathrm{el}(\mathrm{D})} \ , \qquad \cdots \qquad \frac{a \ \varepsilon \ \mathrm{el}(\mathrm{D})}{\mathrm{c}_9(a) \ \varepsilon \ \mathrm{el}(\mathrm{D})} \ .$$

[75] Cf. p. 97.

This makes nineteen constructors in total. There are also nineteen equality rules mirroring the introduction rules. If we abbreviate $\mathrm{c}_i(a)$ by ai, we recover the usual notation for the decimal numbers. For example, with this notation, we have

$$\frac{1 \mathrel{\varepsilon} \mathrm{el}(\mathrm{D})}{10 \mathrel{\varepsilon} \mathrm{el}(\mathrm{D})} \quad \text{and} \quad \frac{10 \mathrel{\varepsilon} \mathrm{el}(\mathrm{D})}{103 \mathrel{\varepsilon} \mathrm{el}(\mathrm{D})} .$$

The sets B, N, 1, $\emptyset$, $A \times B$, $A + B$, and D are inductive in the technical sense defined above. But not all mathematically interesting sets are inductive. I also admit so called coinductive definitions of sets. For a coinductive set, the meaning $a \mathrel{\varepsilon} \mathrm{el}(A)$ is not defined listing the forms which a may have, but, as it were, instrumentally. Examples of coinductive sets are the set $A \to B$ of functions from A to B and the set $\mathrm{S}(A)$ of streams over a fixed set A.[76]

A third kind of definition has formal introduction rules in the same way as the inductive definition, but with a less rigorous demand on equality. The disadvantage of definitions of this kind is that their elimination rules become slightly awkward. An example of such a definition is the formalization of the classical definition of a number as a "multitude composed of units".[77] We have

$$\mathrm{E} \mathrel{\varepsilon} \mathrm{set},$$

where the letter E is used in honour of Euclid.[78] If we denote the unit by 1 and use the plus sign to separate units, we get the introduction rules

$$1 \mathrel{\varepsilon} \mathrm{el}(\mathrm{E}),$$

and

$$\frac{a \mathrel{\varepsilon} \mathrm{el}(\mathrm{E}) \quad b \mathrel{\varepsilon} \mathrm{el}(\mathrm{E})}{a + b \mathrel{\varepsilon} \mathrm{el}(\mathrm{E})} .$$

In particular, zero is not a number in the present sense of the word. The novelty of this definition lies in the treatment of equality; two numbers are not equal *only* if they have equal form and parts, but also when they differ only in the way the terms are associated. That is, in addition to the mandatory equality rules $1 = 1 \mathrel{\varepsilon} \mathrm{el}(\mathrm{E})$ and

$$\frac{a = c \mathrel{\varepsilon} \mathrm{el}(\mathrm{E}) \quad b = d \mathrel{\varepsilon} \mathrm{el}(\mathrm{E})}{a + b = c + d \mathrel{\varepsilon} \mathrm{el}(\mathrm{E})} ,$$

we also have associativity, i.e., the equality rule

$$\frac{a \mathrel{\varepsilon} \mathrm{el}(\mathrm{E}) \quad b \mathrel{\varepsilon} \mathrm{el}(\mathrm{E}) \quad c \mathrel{\varepsilon} \mathrm{el}(\mathrm{E})}{(a + b) + c = a + (b + c) \mathrel{\varepsilon} \mathrm{el}(\mathrm{E})} . \tag{a}$$

In addition, the transitivity rule is an equality rule

$$\frac{a = b \mathrel{\varepsilon} \mathrm{el}(\mathrm{E}) \quad b = c \mathrel{\varepsilon} \mathrm{el}(\mathrm{E})}{a = c \mathrel{\varepsilon} \mathrm{el}(\mathrm{E})} , \tag{b}$$

---

[76] Cf. Ch. V, § 5.

[77] Euclid, *Elementa*, Bk. 7, Def. 2.

[78] Cf. Poincaré, *La Science et l'Hypothèse*, Ch. 1; Aristotle, *Metaph.*, Bk. 13; Kronecker, 'Über den Zahlbegriff', § 1; Bernays, 'Die Philosophie der Mathematik und die Hilbertsche Beweistheorie', § 3; and Frege, *Die Grundlagen der Arithmetik*, §§ 35–44 for critical discussions on this notion of number.

and the converse to associativity is also an equality rule

$$\frac{a \,\varepsilon\, \mathrm{el(E)} \quad b \,\varepsilon\, \mathrm{el(E)} \quad c \,\varepsilon\, \mathrm{el(E)}}{a + (b + c) = (a + b) + c \,\varepsilon\, \mathrm{el(E)}} \,. \tag{c}$$

Now we have to make sure that the equality so defined is reflexive and cancellable. It is clearly reflexive since the introduction rules are mirrored by equality rules. Furthermore, it is easy to make sure that equality is symmetric. Now, since transitivity is already a part of the definition of equality, equality is also cancellable. We may now recognize the formation rule for E as valid.

It should be noted that the equality relation so defined on the set E is decidable in the sense that given two terms a and b, we can, as it were, mechanically find out whether the assertion $a = b \,\varepsilon\, \mathrm{el(E)}$ is demonstrable or not. If both a and b are 1, equality is demonstrable. If one of them is 1, but the other is not, equality is not demonstrable. Thus assume that both a and b have '+' as their outermost form. Now repeatedly rewrite a using (a) and b using (c) and compose the steps by (b), giving the final result

$$a = \cdots = 1 + c = 1 + d = \cdots = b \,\varepsilon\, \mathrm{el(E)}.$$

Next recursively apply the same mechanical procedure to c and d. Eventually the problem of deciding equality between a and b is answered in the positive or in the negative.

This example could be generalized by taking arbitrary elements of a fixed set A instead of the unit 1. In this case we get the set of nonempty finite sequences over the fixed set A. Taking A to be the unit set, we get back a set isomorphic to E. These examples could be further extended by adding the empty sequence or, in the case of E, a zero element. It is left to the reader to work out the details.

To motivate this new kind of definition of equality, consider the distinction between concept and expression. Conceptually we can join multitudes composed of units without making a distinction between multitudes which differ only in the order they are joined together, so all axioms are really justified in virtue of the Euclidean definition of the notion of number as a multitude of units.

This Euclidean definition of number shows clearly that the concept of number is the result of a formal abstraction from the things that are counted.[79] It also shows why one, or the unit, is called the principle of number. The unit, *one*, can be taken in two different, but related, senses: "duplex est unum".[80] The arithmetic unit, which serves as a measure of things counted, the principle of number, presupposes the ontological unit, which is convertible with being. That is, to count, one has to specify *what* to count, i.e., what to count as a being, and it is the unit in this sense of delimitation which *converts*, or goes together, with being in the sense that *ens et unum convertuntur*. The zero also has a metaphysical impact on counting, because, to count, one has first to know where to start, i.e., what to count as nothing, or zero, of the kind of thing counted.[81]

---

[79] Cf. Goodstein, 'The Arabic Numerals, Numbers and the Definition of Counting'.
[80] Aquinas, 'De Potentia', q. 3 a. 16 ad 3. Cf. Aristotle, *Metaph.*, Bk. 5, Ch. 6
[81] Cf. Goodstein, *Recursive number theory*, Intro.

CHAPTER IV

Reference and Computation

In the early history of computation, one finds algorithms like the Babylonian algorithm for the approximation of square roots, Archimedes' algorithm for the approximation of π, Euclid's algorithm for computing the greatest common divisor of two numbers, al-Khwārizmī's algorithms, etc.[1] All these modes of computation, or algorithms, have in common that they produce a result for every conceivable input. In fact, it is safe to assume that the ancients would view an algorithm as erroneous if it did not always produce a result. Dirichlet's much debated pathological function φ was the first function in conflict with this classical notion of algorithm.[2] At that time there was no theory of algorithms and no Church-Turing thesis but, still, the function was criticized for not being *prima facie* computable, meaning that its definition,

$$\varphi(x) = \begin{cases} 1 & \text{if } x \text{ is rational} \\ 0 & \text{otherwise,} \end{cases}$$

does not give sufficient information on how to compute it; this is because there is no systematic way of determining whether a given real number is rational or not. The modern approach to this problem is to treat the notions of function and computation separately. In extensional set theory, a function from one set to another is typically taken to be a subset of their Cartesian product satisfying certain conditions.[3] Intuitionists on the other hand contend that the notion of non-computable function makes little sense; for intuitionists, Dirichlet's φ is in fact not a function at all.[4]

The notions of computation and function, central in intuitionistic type theory, are investigated in the first two sections of the present chapter. The third section explains a systematic way of describing computations, thus anticipating the fourth section where noncanonical sets and elements are defined in terms of computation.

---

[1]Cf. Ifrah, *Räknekonstens kulturhistoria*, Vol. 2, pp. 286, 423–424 (original in French: *Histoire universelle des chiffres*).

[2]Dirichlet, 'Sur la convergence des séries trigonométriques qui servent à représenter une fonction arbitraire entre des limites données', p. 169. In modern terms, φ is the characteristic function of the rational numbers. It is irrelevant to the example whether we let φ take on the values 1 and 0, as Dirichlet did, or the Boolean truth values.

[3]E.g., Krivine, *Introduction to Axiomatic Set Theory*, p. 11.

[4]To be more precise, it is not *prima facie* a function in the sense defined by Bishop and Bridges, *Constructive Analysis*, p. 15. The phrase *prima facie* is used because the method used by intuitionists to show that an alleged function is not to be admitted, viz., to show that it entails a nonconstructive principle, i.e., to give a *weak counterexample*, typically only shows that the function is nonrecursive—not that is noncomputable. The distinction between recursive and computable is explained below.

J.G. Granström, *Treatise on Intuitionistic Type Theory*, Logic, Epistemology, and the Unity of Science 22, DOI 10.1007/978-94-007-1736-7_4,

§ 1. Functions, algorithms, and programs

In this section, I intend to establish preliminary definitions of the notions of computation, function, algorithm, and program. The very notion of computation is abstracted from actual computations performed, or seen performed, for example on paper or on a black-board. Any actual and finished computational process is, since it is seen in its entirety, finite in time; consequently, finiteness is a characteristic of computation. From mathematical practice, it is clear that finiteness is only *in principle*: an expression like

$$1000^{1000^{1000}}$$

is perfectly legitimate in mathematics even though it is computationally unfeasible.[5] Mathematicians agree to call such expressions computable because it is possible to compute their value in finite time; but, as often is the case with things possible, it is possible only in principle. Is then finiteness the only characteristic of computation? I think two more characteristics are necessary, viz., exactness and typing.

By exactness I mean the kind of exactness, or rigor, found in mathematics, in which there are two components. First, that the definitions and proofs are meticulous and worked out in greatest possible detail. Second, a component which is implied by the nature of the subject matter, that the ideas involved are timeless and changeless. It is these two components that give mathematics its unique flavor and which make mathematical exactness an ideal for other sciences.[6]

In the absence of exactness I would no longer speak of computation. This means that when dealing with temporal and contingent being, we sometimes cannot say that the expressions refer to their objects through computation. An example is given by *denoting phrases* like *the most populous city in the world* or *the King of France.*[7] In general, denoting phrases can have any number of possible denotations, including zero.

The third feature of computation, which is easy to overlook, is that even though the value of the computation is not known beforehand, it is always known what type of value to expect: we can have a number valued computation, a Boolean valued computation, etc., but never simply a computation.

A preliminary definition of computation is now given by: a computation is a finite and exact mode of procedure by which an expression refers to an object of a certain type. That is, computation has three characteristics: finiteness, exactness, and typing.

---

[5]Cf. Frege, *Die Grundlagen der Arithmetik*, § 89.

[6]This prominent view of mathematics goes back a long way, cf. Aristotle, *Metaph.*, Bk. 2, Ch. 3, § 3.

[7]The second example is due to Russell, 'On Denoting', p. 479. It is a particularly well chosen example as it is further complicated by France's having two kings after the death of Pepin III in 768. As an aside, I think Russell's treatment of denotation is unnecessarily complicated. A simpler approach is to say that any use of a denoting phrase presupposes that it has a unique referent or denotation. For example, we can speak about the smallest prime number, because its presupposition is fulfilled, but not about the largest prime number, because its presupposition is not fulfilled.

Let us now turn our attention to the notion of algorithm. Despite being a very old notion, all attempts at rigorously defining it are comparatively recent. One of the modern definitions is due to Markov, who defines it as "an exact prescription, defining a computational process, leading from various initial data to the desired result", and furthermore gives the following three characteristic features:

"a) the precision of the prescription, leaving no place to arbitrariness, and its universal comprehensibility—the definiteness of the algorithm;

b) the possibility of starting out with initial data, which may vary within given limits—the generality of the algorithm;

c) the orientation of the algorithm toward obtaining some desired result, which is indeed obtained in the end with proper initial data—the conclusiveness of the algorithm."[8]

Knuth takes finiteness, definiteness, input, output, and effectiveness as characteristics of the notion of algorithm in his highly esteemed work on computer programming, and it is interesting to compare these characteristics with with Markov's definition.[9] Definiteness is common to both definitions, and is discussed below. Its clear that Knuth's input corresponds to Markov's generality: an algorithm takes an input of a certain kind, e.g., an arbitrary number. Markov does not explicitly mention the typing of the output, but since he considers only number valued computations, this is not surprising. Reading Knuth in greater detail, it is clear that what he calls effectiveness is what is called exactness in the above analysis of computation, and what Markov mentions as exactness in his preliminary definition. Finally, what Markov calls conclusiveness is what I have called finiteness, in agreement with Knuth. So there seems to be some consensus as to the meaning of the word algorithm.

The characteristic finiteness is related to the notions of partial and total correctness, as employed in computer science. An "algorithm" is called partially correct if its output, *if any*, is of the correct type; moreover, it is called totally correct if, in addition, it always gives an output. According to our definition, an algorithm is always totally correct in this sense, and this is why the word algorithm was put within quotation marks in the previous sentence. I will use the word program for the notion of algorithm minus finiteness, according to the equation

$$\text{algorithm} = \text{program} + \text{finiteness}.^{10}$$

Still, a program is always partially correct.[11]

---

[8]Markov, 'The theory of algorithms', n. 1.

[9]Knuth, *The Art of Computer Programming*, Ch. 1, § 1.

[10]This, and subsequent equations of this kind, should of course be understood in a vague sense, and not as a suggestion of an arithmetic for concepts.

[11]In a weak type system, one should also make the distinction between type correctness and, as it were, intentional correctness, i.e., that the algorithm or program works as intended, typically according to a specification. For example, if, when asked to provide an algorithm for multiplication, one provides the algorithm for addition, the produced algorithm will be totally correct in the above sense, but obviously not the intended answer. One important property of intuitionistic type theory is that the type system is so expressive that, in most cases, the specification can be made part of the type.

Of Knuth's five characteristics of the notion of algorithm, three are present already for the notion of computation, viz., finiteness, output, and effectiveness (or exactness). The *input* is related to the abstraction by which we get the notion of a function f from the computations $f(a)$ for particular arguments a, or, conversely, to the saturation by which we get a computation $f(a)$ from a function f and an argument a. That is, the characteristic *input* is of a special kind: a function supplied with its argument (input) is a computational expression.

Moreover, I would like to contrast *definiteness*, as a characteristic of the concept of algorithm, to *lawlessness* which intuitionists apply to sequences of numbers and to number theoretic functions in the distinction between law-like, or law-abiding, and lawless sequences of numbers.[12] I propose the following equation:

$$\text{algorithm} = \text{function} + \text{definiteness}.$$

If we take the liberty to view a concept as a sum of its characteristics, this gives the characteristics *input*, *output*, *finiteness*, and *exactness*, for the concept of function.[13] Another word, used in connection with recursion theory, is *recursive*, which I will treat as synonymous with the word *definite*. Thus, a recursive function is simply an algorithm. Moreover, in intuitionistic type theory, the word *method* is often used in connection with the explanation of implication.[14] In this setting, a method is to be understood as a computational expression or as a function, as appropriate.

I have taken the liberty to not include definiteness in the notion of computation. Unfortunately, the word computable is sometimes taken to be synonymous with recursive, leading to unnecessary confusion. Recursion theory, as a subject, was born in 1936 with contributions by Church, Turing, Kleene, Gödel, and Post. During the following years, it became clear that the concept of algorithm on the numbers can be adequately formalized in any of the following three equivalent ways: Church's λ-calculus, Turing's tape machines, and Post's production systems.[15] These computational models should not lead us to abandon the results gained by our investigation of the concepts of computation and algorithm, because these results are conceptually prior to the formalizations.

"The most important discovery in the science of algorithms was undoubtedly the discovery of the general notion of algorithm itself as a new and separate entity. We emphasize that this discovery should not be confused with the discovery of representative computational models (constructed by Turing, Post, Markov, Kolmogorov)... Sometimes it is wrongly believed that the concept of algorithm cannot be satisfactorily understood without certain formal constructions... But we are of the opinion that these constructions were only introduced in order to provide a formal characterization of the in-

[12] Troelstra, *Principles of Intuitionism*, § 5. Of course I mean to identify definite with law-abiding and indefinite with lawless.

[13] The distinction, between function and algorithm, is not novel: a similar distinction was made already by Turing ('On Computable Numbers', p. 232), viz., between automatic machines (algorithms) and choice machines (functions).

[14] Cf. p. 29 of this book, and Martin-Löf, *Intuitionistic Type Theory*, p. 12.

[15] Cf. Church, 'An unsolvable problem of elementary number theory'; Post, 'Formal Reductions of the General Combinatorial Decision Problem'; Turing, 'On Computable Numbers'.

formal concept of algorithm. Thus the concept itself was recognized as existing independently from this formal characterization and as preceding in time. As Gödel indicated,[16] the question whether Turing's definition of the computability of a function is adequate is meaningless unless the notion of computable function is intelligible *a priori*."[17]

Recall the difference between *program*, *function*, and *algorithm*: they have the characteristics *input*, *output*, and *exactness* in common, *function* adding *finiteness*, *program* adding *definiteness*, and *algorithm* adding both. The conceptual priority between these concepts has been disputed: the Russian constructivists take algorithms as their basic concepts while the Dutch intuitionists take function as theirs. In intuitionistic type theory, function is taken as the basic concept, and I think this choice is well-motivated. First because every algorithm is a function; next because the notion of algorithm is easy to define once the notion of function is in place; and finally because it is not clear how the notion of function is to be defined in terms of the notion of algorithm.

In recursion theory, the word computable was originally taken to include both finiteness and definiteness,[18] but in modern recursion theory a computable function is often taken to be what I call a program. It is foundationally problematic to view the notion of program as conceptually prior to algorithm and function. In the modern approach to semantics of programs, the notion of function is assumed to be already in place.[19] If we naïvely try to give meaning to programs by a definition like

$$\mu(p,a) = \begin{cases} b & \text{if } p \text{ with input } a \text{ terminates with value } b, \\ \perp & \text{otherwise,} \end{cases}$$

where $\perp$ is used to indicate that the program p does not terminate,[20] we have to explain what kind of entity μ is. It is well known from recursion theory that, if we allow sufficiently complicated programs, μ cannot be given by an algorithm. There is a metaphysical commitment behind the assumption that there is such a function μ, since, if the above definition of μ is legitimate, it defines a lawless function. If we instead take μ to be a program, we end up with a vicious circle since program is the notion we are trying to define.

§ 2. The concept of function

For an interesting account of the early history of the concept of function, the reader is referred to Youschkevitch's article on the concept of function up to the middle of the 19th century.[21] In this survey, I will concentrate on the concept of function taken in isolation, i.e., without connection to real analysis and the question of continuity. Historically, the concepts of function and continuity were developed side by side, but, conceptually, they are quite distinct.

---

[16] Gödel, 'Über eine bisher noch nicht benütze Erweiterung des finiten Standpunktes' (Reference part of citation).

[17] Uspensky and Semenov, 'What are the gains of the theory of algorithms', § 1.

[18] E.g., in Turing, 'On Computable Numbers', p. 233.

[19] Scott and Strachey, *Toward a Mathematical Semantics for Computer Languages*. Cf. Floyd, 'Assigning meaning to programs'.

[20] As in Backus, 'Can programming be liberated from the von Neumann style? '.

[21] Youschkevitch, 'The concept of function up to the middle of the 19th century'.

The word function was first used in its present sense by Leibniz and further developed in correspondence with Joh. Bernoulli.[22] Later contributions were made by, among others, Euler and Dirichlet. Euler's first definition reads as follows:

"A function of a variable quantity is an analytic expression composed in any way of the variable quantity and numbers or constant quantities."[23]

Euler gives $a + 3z$ and $az - 4z^2$ as examples of functions of the variable z, where the a is a parameter. In the setting of intuitionistic type theory and with reference to our analysis of expressions into form and parts, this definition can be clarified and generalized by saying that a function of a variable x (of a certain kind) is an expression (of a certain kind) built up from the variable x and previously introduced forms.

To construct an example, recall that the logical category 'number' introduced on p. 31 is now to be identified with the logical category el(N), where N is understood as a set in the sense of the descriptive definition on p. 59. The demonstration

$$\dfrac{a : \mathrm{el}(\mathrm{N}) \qquad \dfrac{3 : \mathrm{el}(\mathrm{N}) \quad z : \mathrm{el}(\mathrm{N})}{3 \times z : \mathrm{el}(\mathrm{N})}}{a + (3 \times z) : \mathrm{el}(\mathrm{N})}$$

shows that $a + 3 \times z$ is a function of z in the sense of Euler.

This old-fashioned notion of function of variables is not completely exact since it treats of demonstration completely formally.[24] Still, if A and B are sets, I will write

$$\begin{array}{c} x : \mathrm{el}(A) \\ \vdots \\ f(x) : \mathrm{el}(B) \end{array}$$

for a formal demonstration of that $f(x)$ is an element of the set B from the premiss $x : \mathrm{el}(A)$. Moreover, if $a : \mathrm{el}(A)$, I will write $f(a)$ for the expression $f(x)$ with all relevant occurrences of x replaced by a. Observe that the requirement on form and parts in Definition 3 entails that $f(a)$ and $f(b)$ are equal elements of the set B whenever a and b are equal elements of A.

Euler's first definition of the concept of function is clearly the foundation of Frege's concept of function, which he defines as follows:

"If in an expression, whose content need not be capable of becoming a judgement, a simple or compound sign has one or more occurrences and if we regard that sign as replaceable in all or some of those occurrences by something else (but everywhere by the same thing), then we call the part that remains

---

[22]The first published occurrence of this use of the word function is in Leibniz's seminal paper on calculus 'Nova methodus pro maximis et minimis', but it occurs several years earlier in his unpublished manuscripts.

[23]Author's translation of Euler, *Introductio in Analysin Infinitorum*, p. 4: "Functio quantitatis variabilis, est expressio analytica quomodocunque composita ex illa quantitate variabili, & numeris seu quantitatibus constantibus."

[24]Cf. Table 6 on p. 108.

invariant in the expression a function, and the replaceable part the argument of the function."[25]

The above is Frege's formulation of his so-called abstraction principle which can be viewed as a generalization of Euler's first notion of function from a mathematical language with variables to any language, e.g., instead of using the function $x + x$, we can view *two plus two* as a function of *two*. Another contribution to the concept of function is that he allowed the value of a function to be a proposition, thereby giving rise to the propositional functions ubiquitous in modern logic. An example of a propositional function is

$$x^2 < x + 2,$$

viewed as a function of x. I view this expression strictly as a propositional function, i.e., as an expression in which we can substitute a number for x to get a proposition. In no way is it to be identified with the set of values for which the proposition is true or something similar to that.

Euler's second definition is taken from the preface to the *Institutiones*, and reads as follows:

"Thus when some quantities so depend on other quantities, that if the latter are changed the former undergo change, then the former quantities are called functions of the latter; this definition applies rather widely, and all ways, in which one quantity could be determined by others, are contained in it. If therefore x denotes a variable quantity, then all quantities, which depend upon x in any way, or are determined by it, are called functions of it".[26]

An important step has been taken between these two definitions. This can be seen from Euler's prime example of a function in the second sense, viz., the example with the cannon: [27] when firing a cannon, the distance traveled by the cannonball (l say) is a function of the elevation of the barrel (ϕ say), of the amount of gunpowder (m say), and several other quantities; this would certainly be the case even if no analytic expression $l = f(\phi, m)$ were at hand. Thus, Euler's second definition is a real development of his first. Of course, the difference between variables and parameters is not so clear-cut in examples taken from physics: "one man's constant is another man's variable".[28] But this is a problem of experimental methodology, and does not affect the notion of function considered in itself. A function in Euler's second sense is conceived as something with foundation in some reality, e.g., in a physical law, in an algorithm, or, to anticipate an example given below, in the tossing of a coin.

Consider now the functions:

$$\mathrm{f}(z) = 3z - 4z^2$$

---

[25]Frege, *Begriffsschrift*, § 9 (trans. Bauer-Mengelberg in van Heijenoort, *From Frege to Gödel*, p. 22).

[26]Author's translation of Euler, *Institutiones Calculi Differentialis*, p. vi: "Quae autem quantitates hoc modo ab aliis pendent, ut his mutatis etiam ipsae mutationes subeant, eae harum functiones appellari solent; quae denominatio latissime patet, atque omnes modos, quibus una quantitas per alias determinari potest, in se complectitur. Si igitur x denotet quantitatem variabilem, omnes quantitates, quae utcunque ab x pendent, seu per eam determinantur, eius functiones vocantur".

[27]Loc. cit.

[28]Perlis, 'Epigrams on Programming', n. 1.

and

$$\mathrm{p}(x) = (x^2 < x + 2),$$

where $\mathrm{f}(z)$ is a function of the variable z in the sense of Euler, and $\mathrm{p}(x)$ is a propositional function of the variable x in the sense of Frege. This notation, together with Euler's second definition of the concept of function, hints at the fact that it might be the forms f and p that are the functions, and not the analytic expressions; this leads to the modern concept of function, or the *black-box* notion of function. One of the earliest definitions of this concept is due to Peano:

"Let φ be a sign, or aggregate of signs, such that if x is an entity of class S, the sign combination φx determines a new entity; we also suppose an equality defined between the entities φx; and if x and y are entities of the class S, and $x = y$, we suppose that it can be deduced that $\varphi x = \varphi y$. Then the sign φ is said to be *a prefix sign for a function in the class* S, and we write $\varphi \,\varepsilon\, \mathrm{F}`S$."[29]

A similar definition is due to Dedekind:

"By a *transformation* φ of a system S we understand a law according to which to every determinate element s of S there *belongs* a determinate thing which is called the *transform* of s and denoted by $\varphi(s)$."[30]

Consider the double function,

$$\mathrm{dbl}(x) = x + x,$$

and let us investigate what Dedekind and Peano have to say about it applied to s(0). According to Dedekind, dbl is a transformation of the system N, s(0) is a determinate element of N, and to it belongs the determinate thing s(s(0)), which is called the transform of s(0) and is denoted dbl(s(0)); according to Peano, dbl(s(0)) is a writ determining a new element of N, and *the sign* dbl is a prefix sign for a function in class N. In neither case is any information given about in what sense dbl(s(0)) *determines a new entity*, or how *a determinate thing belongs* to it. To see why this is problematic, the analogy of cooking is helpful. The function dbl is like a recipe, and its argument s(0) like the ingredients. Writing dbl(s(0)) is like placing the recipe next to the ingredients, and this does not make a meal. The cooking of the meal corresponds to the *computation* of dbl(s(0)). Using the distinction between canonical and noncanonical terms, dbl(s(0)) is a noncanonical number and that its value s(s(0)) is a canonical number: the former refers to the latter. In general, mathematical expressions refer to their value by computation.

---

[29]Author's translation of Peano, *Arithmetices Principia Nova Methodo Exposita*, Log. Not., n. 6, p. 30: "Sit φ signum, sive signorum aggregatus, ita ut si x est ens classis s, scriptura φx novum indicet ens; supponimus quoque aequalitatem inter entia φx definitam; et si x et y sunt entia classis s, et est $x = y$, supponimus deduci posse $\varphi x = \varphi y$. Tunc signum φ dicitur esse *functionis praesignum in classi* s, et scribemus $\varphi \,\varepsilon\, \mathrm{F}`s$." (In the translation I have capitalized the s to make it conform with Dedekind's notation.)

[30]Dedekind, 'The Nature and Meaning of Numbers', § 2. The original German reads: "Unter einer *Abbildung* φ eines Systems S wird ein Gesetz verstanden, nach welchem zu jedem bestimmten Element s von S ein bestimmtes Ding *gehört*, welches das *Bild* von s heißt und mit $\varphi(s)$ bezeichnet wird."

To make this point clearer, we consider a very short expression for a very complicated computation. The Goldbach conjecture is usually formulated as: every even number can be written as a sum of at most two prime numbers. Define the function gold from numbers to Booleans, by

$$\mathrm{gold}(x) = \begin{cases} 1 & \text{if the Goldbach conjecture holds up to } x \\ 0 & \text{otherwise;} \end{cases}$$

it is clear that this function can be computed by an algorithm, and it has been verified that it is constantly 1 up to very large values of x. The full Goldbach conjecture is now that the function gold is constantly equal to 1. Furthermore, let

$$\mathrm{ack}(x) = \varphi(x, x, x),$$

where φ is the Ackermann function.[31] Now consider the expression

$$\mathrm{gold}(\mathrm{ack}(9)).$$

This expression denotes a Boolean but, I think, nobody knows which Boolean. That is, it is a noncanonical Boolean without known value. This example shows that one can understand a mathematical expression without knowing its value. In normal mathematical practice, one is used to see straight through the expressions, viewing them as everywhere replaced by their value; though it has to be remembered that this practice is legitimated only by finiteness of computations and, furthermore, that this finiteness is only in principle. In the analysis of the notions of denotation and reference this replacement is no longer possible, and one has to get used to a new way of thinking.

The first and most basic sense in which f, p, dbl, ack, and gold are functions is in the sense of Peano. I will call a unary form f a function in the sense of Peano, or a functional form, if the inference rules

$$\frac{a : \mathrm{el}(A)}{f(a) : \mathrm{el}(B)}$$

and

$$\frac{a = b : \mathrm{el}(A)}{f(a) = f(b) : \mathrm{el}(B)}$$

are valid. The unary forms f, p, dbl, ack, and gold are functional forms in this sense. A functional form can be defined from an analytic expression by a nominal definition of the form

$$\overbrace{\mathit{form}}^{\text{functional form}}(z) = \underbrace{a + 3 \times z}_{\text{analytic expression}} : \mathrm{el}(\mathrm{N}).$$

The definitions of the forms dbl and ack provide two examples. Observe that we cannot speak about the unary form f inside the language of intuitionistic type theory, so it is a metalinguistic notion, i.e., functional forms are spoken about only in the metalanguage.

[31] The Ackermann function, defined in 'Zum Hilbertschen Aufbau der reellen Zahlen', p. 120, has the property that $\varphi(a, b, 0) = a + b$, $\varphi(a, b, 1) = a \cdot b$, $\varphi(a, b, 2) = a^b$, etc., whence $\mathrm{ack}(0) = 0$, $\mathrm{ack}(1) = 1$, $\mathrm{ack}(2) = 4$, $\mathrm{ack}(3) = 3^{27}$ and ack(4) is too large to write in this footnote using standard mathematical notation. The point is that ack is a recursive function which grows very quickly.

§ 3. A formalization of computation

Even though I recognize the notion of computation as being independent from any formalization, it is still valuable to have a formalism in which to express computations. Here I have chosen to adopt a formalism germane to what Plotkin called structural operational semantics.[32] To avoid complexity, I give some examples of number valued computations before treating the matter in full detail. The notation

$$a \Rightarrow b \;\varepsilon\; \mathrm{el(N)}$$

is read: computing the term a gives the value b which is a canonical element of N.[33] The notion of canonical number is already understood, and this notation is introduced to make the notion of noncanonical number exact.

The predecessor function is defined by the computation rules

$$\mathrm{pred}[0] \Rightarrow 0 \;\varepsilon\; \mathrm{el(N)}$$

and

$$\mathrm{pred}[\mathrm{s}(a)] \Rightarrow a \;\varepsilon\; \mathrm{el(N)},$$

and the double function is defined by the computation rules

$$\mathrm{dbl}[0] \Rightarrow 0 \;\varepsilon\; \mathrm{el(N)}$$

and

$$\frac{\mathrm{dbl}[a] \Rightarrow b \;\varepsilon\; \mathrm{el(N)}}{\mathrm{dbl}[\mathrm{s}(a)] \Rightarrow \mathrm{s}(\mathrm{s}(b)) \;\varepsilon\; \mathrm{el(N)}}\,.$$

Both pred and dbl are functional forms in the sense defined above.

A computation rule either has the form of an assertion or the form of an inference rule, but it is read bottom-up. Recall that the epsilon sign is used for canonical elements of a set and the colon sign for noncanonical elements. The above computation rules immediately justify the inference rules

$$\frac{a \;\varepsilon\; \mathrm{el(N)}}{\mathrm{pred}[a] : \mathrm{el(N)}}$$

and

$$\frac{a \;\varepsilon\; \mathrm{el(N)}}{\mathrm{dbl}[a] : \mathrm{el(N)}}\,,$$

which make it possible to build noncanonical terms from canonical terms. We often want to build noncanonical terms from other noncanonical terms. If the parts of the predecessor or double forms are noncanonical, they are evaluated first, that is eagerly, using the computation rules

$$\frac{a \Rightarrow b \;\varepsilon\; \mathrm{el(N)} \quad \mathrm{pred}[b] \Rightarrow c \;\varepsilon\; \mathrm{el(N)}}{\mathrm{pred}(a) \Rightarrow c \;\varepsilon\; \mathrm{el(N)}}$$

[32] Plotkin, *A Structural Approach to Operational Semantics*; cf. Clement et al., *Natural Semantics on the Computer*.

[33] It is important to remember that computation, first and foremost, is an *act*. Just as Hamlet really exists only when being played, Euclid's algorithm really exists only when being performed. "To be dependent on self-presentation belongs to what it is", as Gadamer writes about a work of art (*Truth and Method*, Ch. 2, § 1.b).

and

$$\frac{a \Rightarrow b \;\varepsilon\; \mathrm{el(N)} \quad \mathrm{dbl}[b] \Rightarrow c \;\varepsilon\; \mathrm{el(N)}}{\mathrm{dbl}(a) \Rightarrow c \;\varepsilon\; \mathrm{el(N)}}\,.$$

In virtue of these computation rules, the inference rules

$$\frac{a : \mathrm{el(N)}}{\mathrm{pred}(a) : \mathrm{el(N)}}$$

and

$$\frac{a : \mathrm{el(N)}}{\mathrm{dbl}(a) : \mathrm{el(N)}}$$

are recognized as valid. As a general rule, square brackets are filled with canonical elements and round brackets with noncanonical elements.[34] This way of computing complex mathematical expressions is called eager, or applicative-order, evaluation. Note that pred and dbl are now functional forms in the sense defined above.

Now add the computation rules

$$0 \Rightarrow 0 \;\varepsilon\; \mathrm{el(N)}$$

and

$$\frac{a \Rightarrow b \;\varepsilon\; \mathrm{el(N)}}{\mathrm{s}(a) \Rightarrow \mathrm{s}(b) \;\varepsilon\; \mathrm{el(N)}}\,,$$

by means of which the assertion

$$0 : \mathrm{el(N)}$$

and the inference rule

$$\frac{a : \mathrm{el(N)}}{\mathrm{s}(a) : \mathrm{el(N)}}$$

are recognized as valid. Using these axioms and inference rules, it is now easy to demonstrate the assertion

$$\mathrm{dbl(dbl(s(0))) : el(N)}.$$

According to what this assertion means, the subject must have a value which is a canonical number. Indeed, we have

$$\frac{\dfrac{\dfrac{0 \Rightarrow 0}{\mathrm{s}(0) \Rightarrow \mathrm{s}(0)} \quad \dfrac{\mathrm{dbl}[0] \Rightarrow 0}{\mathrm{dbl}[\mathrm{s}(0)] \Rightarrow \mathrm{s}(\mathrm{s}(0))}}{\mathrm{dbl}(\mathrm{s}(0)) \Rightarrow \mathrm{s}(\mathrm{s}(0))} \quad \dfrac{\dfrac{\mathrm{dbl}[0] \Rightarrow 0}{\mathrm{dbl}[\mathrm{s}(0)] \Rightarrow \mathrm{s}(\mathrm{s}(0))}}{\mathrm{dbl}[\mathrm{s}(\mathrm{s}(0))] \Rightarrow \mathrm{s}(\mathrm{s}(\mathrm{s}(\mathrm{s}(0))))}}{\mathrm{dbl}(\mathrm{dbl}(\mathrm{s}(0))) \Rightarrow \mathrm{s}(\mathrm{s}(\mathrm{s}(\mathrm{s}(0))))}\,,$$

where ε el(N) has been left out everywhere, since all computations are of numbers anyway. This shows the connection between computation rules and noncanonical elements.

[34]This way of writing expression raises the question of what the forms of dbl(a) and dbl[a] are. My answer is that the form of dbl[a] is dbl[·] and that the form of dbl(a) is dbl(·); though, in the latter case, the form is also spelled simply dbl. Other explanations are also possible: the important thing is that each form comes in two "flavors", one with canonical parts and one with noncanonical parts. It could be objected that, for consistency, one should use square brackets also for constructors, but this is not necessary since the notation is unambiguous anyway.

A second way of computing expressions is by lazy evaluation. The difference lies in whether or not the parts of an expression are evaluated before the whole expression. Other terminologies for the dichotomy are strict evaluation vs. non-strict evaluation and applicative-order evaluation vs. normal-order evaluation.[35] That is, we have the equalities

$$\text{eager} = \text{applicative-order} = \text{strict},$$

and

$$\text{lazy} = \text{normal-order} = \text{non-strict}.$$

There are subtle differences between the different terminologies, and inconsistencies in their use, but these differences show up only when dealing with the evaluation of programs, as opposed to functions or algorithms, and are of no concern at the moment.[36]

The example of the double function will now be repeated with lazy evaluation. The set of lazy numbers, which I will write N_ℓ, has the introduction rules

$$0 \;\varepsilon\; \mathrm{el}(\mathrm{N}_\ell)$$

and

$$\frac{a : \mathrm{el}(\mathrm{N}_\ell)}{\mathrm{s}(a) \;\varepsilon\; \mathrm{el}(\mathrm{N}_\ell)},$$

and the normal equality rules. That is, the difference between N and N_ℓ is that the a in $\mathrm{s}(a)$ is canonical in the former case but noncanonical in the latter case. The successor form has the computation rule $\mathrm{s}(a) \Rightarrow \mathrm{s}(a) \;\varepsilon\; \mathrm{el}(\mathrm{N}_\ell)$, and, as before, we have $0 \Rightarrow 0 \;\varepsilon\; \mathrm{el}(\mathrm{N}_\ell)$, whence the assertion

$$0 : \mathrm{el}(\mathrm{N}_\ell)$$

and the inference rule

$$\frac{a : \mathrm{el}(\mathrm{N}_\ell)}{\mathrm{s}(a) : \mathrm{el}(\mathrm{N}_\ell)}$$

are valid. The double form has the computation rules

$$\mathrm{dbl}[0] \Rightarrow 0 \;\varepsilon\; \mathrm{el}(\mathrm{N}_\ell)$$

and

$$\mathrm{dbl}[\mathrm{s}(a)] \Rightarrow \mathrm{s}(\mathrm{s}(\mathrm{dbl}(a))) \;\varepsilon\; \mathrm{el}(\mathrm{N}_\ell),$$

and the corresponding inference rule

$$\frac{a \;\varepsilon\; \mathrm{el}(\mathrm{N}_\ell)}{\mathrm{dbl}[a] : \mathrm{el}(\mathrm{N}_\ell)}.$$

The double function applied to a noncanonical element is evaluated according to the computation rule

$$\frac{a \Rightarrow b \;\varepsilon\; \mathrm{el}(\mathrm{N}_\ell) \quad \mathrm{dbl}[b] \Rightarrow c \;\varepsilon\; \mathrm{el}(\mathrm{N}_\ell)}{\mathrm{dbl}(a) \Rightarrow c \;\varepsilon\; \mathrm{el}(\mathrm{N}_\ell)},$$

[35] Cf. Abelson and Sussman, *Structure and Interpretation of Computer Programs*, § 1.1.5, and § 4.2.

[36] When dealing with evaluation of programs, the order (e.g. left to right) in which the parts of a form are evaluated is also of concern. Moreover, there are different strategies of implementation, such as call by value and call by reference for eager evaluation, and call by name and call by need for lazy evaluation. But, again, this need not concern us here.

by means of which we recognize the inference rule

$$\frac{a : \mathrm{el}(\mathrm{N}_\ell)}{\mathrm{dbl}(a) : \mathrm{el}(\mathrm{N}_\ell)} .$$

As above, we have

$$\mathrm{dbl}(\mathrm{dbl}(\mathrm{s}(0))) \; \varepsilon \; \mathrm{el}(\mathrm{N}_\ell),$$

and the computation trace for this term with lazy evaluation becomes

$$\frac{\dfrac{\begin{array}{l}\mathrm{s}(0) \Rightarrow \mathrm{s}(0)\\ \mathrm{dbl}[\mathrm{s}(0)] \Rightarrow \mathrm{s}(\mathrm{s}(\mathrm{dbl}(0)))\end{array}}{\mathrm{dbl}(\mathrm{s}(0)) \Rightarrow \mathrm{s}(\mathrm{s}(\mathrm{dbl}(0)))} \quad \mathrm{dbl}[\mathrm{s}(\mathrm{s}(\mathrm{dbl}(0)))] \Rightarrow \mathrm{s}(\mathrm{s}(\mathrm{dbl}(\mathrm{s}(\mathrm{dbl}(0)))))}{\mathrm{dbl}(\mathrm{dbl}(\mathrm{s}(0))) \Rightarrow \mathrm{s}(\mathrm{s}(\mathrm{dbl}(\mathrm{s}(\mathrm{dbl}(0)))))} ,$$

where again ε el(N_ℓ) has been left out everywhere.[37] It is instructive to compare this with the eager evaluation of the same term. The lazy evaluator stops at the term s(s(dbl(s(dbl(0))))), because it is already in canonical form, which in this case means either the form 0, or the form s(a), where a is a noncanonical number, while the eager evaluator continues until it reaches s(s(s(s(0)))).

There are at least two benefits of eager evaluation. The first is practical, and relates to the implementability of the computations on a computer: lazy evaluation is more difficult to implement effectively (at least this seems to be the opinion of most computer scientists). The second benefit of eager evaluation is that a canonical expression is easier to analyse if its parts are also canonical, e.g., s(s(s(s(0)))) vs. s(s(dbl(s(dbl(0))))). There are also benefits of lazy evaluation, such as it being simpler from a formal point of view.

The present formulation of intuitionistic type theory supports both eager and lazy evaluation within the same formalism, i.e., we do not have to chose between N and N_ℓ, but can have both. This is in contrast to earlier presentations of intuitionistic type theory, where only lazy evaluation is supported. To allow both modes of computation is of particular interest in computer science since it makes it possible to experiment with hybrid systems.

Recall the three characteristics of computation, viz., finiteness, exactness, and typing. A computation rule must always yield finite computations. This is an external condition—nothing prevents us from writing down a rule which does not yield finite computations, but, by definition, it would not be a computation rule. Moreover, a term has a unique value, i.e., exactness of computation. This could be formulated as

$$\frac{a \Rightarrow b \; \varepsilon \; \mathrm{el}(\mathrm{N}) \quad a \Rightarrow c \; \varepsilon \; \mathrm{el}(\mathrm{N})}{b \equiv c} ,$$

where $b \equiv c$ means that b and c are the same term, as long as it is understood that the conclusion is strictly speaking not an assertion. Exactness is also an external condition. Furthermore, the computation rules are typed, whence any a for which $a \Rightarrow b \; \varepsilon$ el(N) for some canonical number b is indeed a computational expression.

[37]The two premisses on the upper left have been written one above the other for typographical reasons—no change in meaning is intended.

In fact, according to what we have seen this far, it is a definite computation. What about indefinite computations, or computations involving choice? This mode of computation is involved when I define a number by saying that I will make up my mind as to its value when you ask me to compute it the first time. Even if a computation may be indefinite in this sense, it is still exact, in particular, it is not subject to change. It makes perfect sense to define an element

$$\text{toss} : \text{el}(\text{N})$$

stating that its value is to be determined by tossing a coin, or in some other way involving a choice, *when* toss *is evaluated.* Such an element, toss, can be said to have two computation rules: first

$$\frac{\text{toss is not determined} \quad \text{determine toss to some } c \, \varepsilon \, \text{el}(\text{N})}{\text{toss} \Rightarrow c \, \varepsilon \, \text{el}(\text{N})},$$

corresponding to the tossing of the coin, and next

$$\frac{\text{toss is determined to } c}{\text{toss} \Rightarrow c \, \varepsilon \, \text{el}(\text{N})},$$

making the computation of toss exact. The element toss of N starts out as not determined, and, when evaluated the first time, only the first computation rule is applicable; an arbitrary choice of c is somehow made by the computer, e.g., by tossing a coin, and toss becomes determined; on subsequent computations, only the second inference rule is applicable. The same principle can be extended to inference rules of the kind

$$\frac{a \, \varepsilon \, \text{el}(\text{N})}{\text{rnd}[a] : \text{el}(\text{N})},$$

recognized in virtue of the computation rules

$$\frac{\text{rnd}[a] \text{ is not determined} \quad \text{determine rnd}[a] \text{ to some } c \, \varepsilon \, \text{el}(\text{N})}{\text{rnd}[a] \Rightarrow c \, \varepsilon \, \text{el}(\text{N})},$$

and

$$\frac{\text{rnd}[a] \text{ is determined to } c}{\text{rnd}[a] \Rightarrow c \, \varepsilon \, \text{el}(\text{N})}.$$

If we add the inference rule

$$\frac{a : \text{el}(\text{N})}{\text{rnd}(a) : \text{el}(\text{N})},$$

recognized in virtue of the computation rule

$$\frac{a \Rightarrow b \, \varepsilon \, \text{el}(\text{N}) \quad \text{rnd}[b] \Rightarrow c \, \varepsilon \, \text{el}(\text{N})}{\text{rnd}(a) \Rightarrow c \, \varepsilon \, \text{el}(\text{N})},$$

the form rnd is completely on a par with the form dbl, and can be used in any setting where the latter can be used. For example, with the machinery introduced in the next chapter, we can make sense of expressions like

$$\sum_{x=0}^{100} \text{rnd}(x).$$

A peculiarity of forms like toss and rnd is that they have a kind of identity. There can be two *different* forms with the *same* verbatim definition.[38] Put differently, my random function need not be the same as yours.

Since computation involves finiteness, an assertion of the form $a : \mathrm{el}(\mathrm{N})$ has a commissive aspect to it.[39] If you have such a term a, and if somebody, which I will call the computer,[40] is prepared to compute it, you can issue the directive

$$\text{Compute } a : \mathrm{el}(\mathrm{N})\,!$$

with the presupposition $a : \mathrm{el}(\mathrm{N})$, and, as result, get a canonical term b, such that $a \Rightarrow b \;\varepsilon\; \mathrm{el}(\mathrm{N})$. The above is reflected in normal mathematical practice when the signs for addition and multiplication are taken to mean that the two terms should be added or multiplied.[41] This is why exercise books in mathematics can get away with presenting mathematical expressions as if they were exercises.

In the above examples I have only used elements of N, that is, numbers, but the same principles apply to elements of any set. Along these lines, I will now proceed to define the form of assertion $a : \mathrm{el}(A)$ with the same exactness as I have previously defined the form of assertion $a \;\varepsilon\; \mathrm{el}(A)$.

§ 4. Noncanonical sets and elements

Now that the relation between computation and reference is explained, it is time to make the descriptive definition of the notion of set (p. 59) completely rigorous. A large portion of the work is already done in the definition of the notion of canonical set (p. 64), and what is missing is, as explained above, the notion of computation. Because this section deals with noncanonical sets and elements, we will shift our terminology and drop the prefix "noncanonical", instead we will use the prefix "canonical" when referring to a canonical set or element. The difference between what we do here and what we did above for the numbers is that here we allow for computation to take place also on the right-hand side of the copula, i.e., we consider not only canonical sets, like the numbers, but also sets which are not canonical, examples of which will be provided later. First, we have the definition of a noncanonical set.

Definition 6. *That A is a noncanonical set, abbreviated A : set, means that the value of A is a canonical set. That the noncanonical set A has the canonical set B as value, abbreviated $A \Rightarrow B \;\varepsilon$ set, is a form of assertion defined by the computation rules which have a conclusion of this form.*

Recall that the colon is used as copula in assertions involving noncanonical objects and the epsilon as copula in assertions involving canonical objects. That A is a set means that $A \Rightarrow B \;\varepsilon$ set for some canonical set B. Moreover,

[38]The technique used to incorporate randomness in computation is germane to the memo technique in computer science, introduced by Michie, 'Memo Functions and Machine Learning'.

[39]The definition of commissive and directive speech acts is found on p. 23.

[40]It does not matter if the computer is a man or a machine as long as he is capable of following the instructions expressed by the computation rules.

[41]Cf. Euler, *Elements of Algebra*, Ch. 1, § 2, n. 8: "$5+3$ signifies that we must add 3 to the number 5".

$A \Rightarrow B$ ε set presupposes that A is a set and that B is a canonical set. This seem like a circular definition, but the situation is analogous to the relation between A ε set and a ε el(A). That A ε set means, among other things, that a ε el(A) is defined, and a ε el(A) presupposes that A ε set. So, just as a canonical set is understood together with its canonical elements, a noncanonical set is understood together with its computation rules.

As we saw above for the numbers, the form of assertion A : set has a commissive aspect to it, i.e., a promise to compute the value of A when demanded to do so. Consider the following sequence of speech acts,

$$\begin{cases} A : \text{set} & \text{(commissive)} \\ \text{Compute } A : \text{set!} & \text{(directive)} \\ \quad\vdots & \\ A \Rightarrow B \;\varepsilon\; \text{set} & \text{(assertive).} \end{cases}$$

Is not A : set in fact a commissive sentence? Not completely: that one recognizes something as possible in principle does not mean that one commits oneself to do it; put differently, to know that A : set is to know that it is, in principle, possible to compute the value of A, not actually to be prepared to do it.[42] In a rigorous study of commissive speech acts, I think that the commissive that c is prepared to compute a : el(A) should have as a presupposition that c knows that a : el(A). In the case when the computer c is a machine, the user imposes on it to compute what he likes—the computer does not have a say in the matter.

As indicated by the definition, the form of assertion $A \Rightarrow B$ ε set is to be understood completely formally. This means that its meaning is determined by the inference rules having a conclusion of this form, of course under the usual conditions of exactness and finiteness.

Exactness means that it is illicit to give two different values to the same set. For example, if the computation rule

$$A \Rightarrow \mathrm{N} \;\varepsilon\; \text{set}$$

is introduced, so that A is a noncanonical set having the canonical set of numbers as value, it would be illicit to also introduce the computation rule

$$A \Rightarrow \emptyset \;\varepsilon\; \text{set},$$

because N and ∅ are distinct canonical sets. I require both canonical and noncanonical terms to be univocal, or unambiguous, in their respective logical category; i.e., the same term cannot denote two different sets, but may well have different meanings in different logical categories.[43] I will call an expression that has different meanings in different logical categories *polymorphic*. Note that different occurrences of a polymorphic expression can stand for different

[42]Cf. the example with gold(ack(9)) on p. 85.

[43]For example, the unit set is spelled 1 and the same symbol is used for an element of the set E; furthermore, the empty set could be spelled 0 without ambiguity. I have also used the same notation, 0 and s(a), for the elements of the two sets N and N_ℓ; this is also harmless as el(N) and el(N_ℓ) are distinct logical categories; similarly, the Boolean values are called 0 and 1, the zero is the same as in N and N_ℓ, and the one is the same as in E.

terms even in the same assertion. Exactness could be expounded by the pseudo inference rule

$$\frac{A \Rightarrow B \ \varepsilon \ \text{set} \quad A \Rightarrow C \ \varepsilon \ \text{set}}{B \equiv C},$$

if we remember that the conclusion, which means that B and C are the same term, is, strictly speaking, not an assertion.

At this point a choice has to be made: either to introduce the notion of noncanonical element of a noncanonical set or the notion of equal noncanonical sets. The order cannot be determined by appeal to conceptual priority. I have chosen to treat the notion of equality between noncanonical sets first simply because it is the easier of the two.

Definition 7. *That A and B are equal noncanonical sets, abbreviated $A = B : \text{set}$, means that their values are equal canonical sets.*

Of course, that A and B are equal sets presupposes that A is a set and that B is a set. The inference rule

$$\frac{A = B : \text{set} \quad A \Rightarrow C \ \varepsilon \ \text{set} \quad B \Rightarrow D \ \varepsilon \ \text{set}}{C = D \ \varepsilon \ \text{set}} \tag{D4.1}$$

is meaning determining for the form of assertion $A = B : \text{set}$. Just as equality between canonical sets is reflexive and cancellable, so is equality between noncanonical sets, i.e., the inference rules

$$\frac{A : \text{set}}{A = A : \text{set}} \tag{J4.1}$$

and

$$\frac{A = C : \text{set} \quad B = C : \text{set}}{A = B : \text{set}} \tag{J4.2}$$

are both valid.

Justification of (J4.1). Let the premiss $A : \text{set}$ be given. This means that $A \Rightarrow B \ \varepsilon \ \text{set}$ for some $B \ \varepsilon \ \text{set}$. By (J3.2), $B = B \ \varepsilon \ \text{set}$, which, since the value of A is uniquely determined, suffices to show that $A = A : \text{set}$. This completes the justification.

Justification of (J4.2). First, let the premisses $A = C : \text{set}$ and $B = C : \text{set}$ be given. Their presuppositions mean that $A \Rightarrow A_0 \ \varepsilon \ \text{set}$, $B \Rightarrow B_0 \ \varepsilon \ \text{set}$, and $C \Rightarrow C_0 \ \varepsilon \ \text{set}$, for some canonical sets A_0, B_0, and C_0. The conclusion means that $A_0 = B_0 \ \varepsilon \ \text{set}$. Reason as follows: $A = C : \text{set}$, therefore $A_0 = C_0 \ \varepsilon \ \text{set}$, by (D4.1), since the values of A and C are as given; $B = C : \text{set}$, therefore, again by (D4.1), $B_0 = C_0 \ \varepsilon \ \text{set}$; therefore $A_0 = B_0 \ \varepsilon \ \text{set}$, by (J3.1), which was to be demonstrated. Note that we have implicitly used uniqueness of value in taking C_0 as the value of C in both places. This completes the justification.

Along familiar lines, symmetry and transitivity can be demonstrated for equality between noncanonical sets, i.e.,

$$\frac{A = B : \text{set}}{B = A : \text{set}} \tag{M4.1}$$

and

$$\frac{A = B : \text{set} \quad B = C : \text{set}}{A = C : \text{set}} \tag{M4.2}$$

are mediate inference rules.

The definition of the notion of noncanonical element of a noncanonical set is more complicated than that of noncanonical set, because we also have to allow for computation to take place on the right-hand side of the copula.

Definition 8. *That a is a noncanonical element of the noncanonical set A, abbreviated $a : \mathrm{el}(A)$, means that the value of a is a canonical element of the canonical set which is the value of A. That the noncanonical element a of the set A has the canonical element b of the canonical set B as value, abbreviated $\mathrm{el}(A) : a \Rightarrow b \; \varepsilon \; \mathrm{el}(B)$, is a form of assertion defined by the inference rules which have a conclusion of this form.*

That a is an element of the set A presupposes that A is a set, and the form of assertion $\mathrm{el}(A) : a \Rightarrow b \; \varepsilon \; \mathrm{el}(B)$ presupposes that $A \Rightarrow B \; \varepsilon$ set, that $a : \mathrm{el}(A)$, and that $b \; \varepsilon \; \mathrm{el}(B)$. Just as for sets, the form of assertion $\mathrm{el}(A) : a \Rightarrow b \; \varepsilon \; \mathrm{el}(B)$ is completely formal: the comments on the corresponding form of assertion for sets apply also here. That the value of a noncanonical element is unique, i.e., exactness, could be expounded by the pseudo inference rule

$$\frac{\mathrm{el}(A) : a \Rightarrow b \; \varepsilon \; \mathrm{el}(B) \quad \mathrm{el}(A) : a \Rightarrow c \; \varepsilon \; \mathrm{el}(B)}{b \equiv c} .$$

The assertion $a : \mathrm{el}(A)$ has a commissive aspect to it but, to compute a, we first have to compute the set to which it belongs, giving us the following sequence of speech acts:

$$\begin{cases} a : \mathrm{el}(A) & \text{(commissive)} \\ \text{Compute } a : \mathrm{el}(A)! & \text{(directive)} \\ \quad \vdots & \\ A \Rightarrow B \; \varepsilon \; \text{set} & \text{(assertive)} \\ \quad \vdots & \\ \mathrm{el}(A) : a \Rightarrow b \; \varepsilon \; \mathrm{el}(B) & \text{(assertive).} \end{cases}$$

The fourth and last form of assertion is equality between elements of a set.

Definition 9. *That a and b are equal noncanonical elements of the set A, abbreviated $a = b : \mathrm{el}(A)$, means that their values are equal canonical elements of the canonical set which is the value of A.*

That a and b are equal elements of A presupposes that a is an element of A and that b is an element of A, which in turn presuppose that A is a set. The following inference rule is immediately evident from this definition,

$$\frac{a = b : \mathrm{el}(A) \quad \mathrm{el}(A) : a \Rightarrow c \; \varepsilon \; \mathrm{el}(C) \quad \mathrm{el}(A) : b \Rightarrow d \; \varepsilon \; \mathrm{el}(C)}{c = d \; \varepsilon \; \mathrm{el}(C)} , \tag{D4.2}$$

and, in addition, completely determines the meaning of the form of assertion $a = b : \mathrm{el}(A)$.

As expected from any sound notion of equality, equality between elements of a set A is reflexive

$$\frac{a : \mathrm{el}(A)}{a = a : \mathrm{el}(A)} \tag{J4.3}$$

and cancellable

$$\frac{a = c : \mathrm{el}(A) \quad b = c : \mathrm{el}(A)}{a = b : \mathrm{el}(A)} . \tag{J4.4}$$

Justification of (J4.3). Let the premiss $a : \mathrm{el}(A)$ be given. This means that $\mathrm{el}(A) : a \Rightarrow b \,\varepsilon\, \mathrm{el}(B)$ for some $b \,\varepsilon\, \mathrm{el}(B)$, where B is the value of A. By (D3.1), $b = b \,\varepsilon\, \mathrm{el}(B)$, which, because the value of a is uniquely determined, suffices to justify that $a = a : \mathrm{el}(A)$.

Justification of (J4.4). Let the premisses $a = c : \mathrm{el}(A)$ and $b = c : \mathrm{el}(A)$ be given. Their presuppositions mean that $\mathrm{el}(A) : a \Rightarrow a_0 \,\varepsilon\, \mathrm{el}(B)$, that $\mathrm{el}(A) : b \Rightarrow b_0 \,\varepsilon\, \mathrm{el}(B)$, and that $\mathrm{el}(A) : c \Rightarrow c_0 \,\varepsilon\, \mathrm{el}(B)$. That $a = b : \mathrm{el}(A)$ now means that $a_0 = b_0 \,\varepsilon\, \mathrm{el}(B)$. Using the premisses given and (D4.2), we get $a_0 = c_0 \,\varepsilon\, \mathrm{el}(B)$ and $b_0 = c_0 \,\varepsilon\, \mathrm{el}(B)$. The justification is completed by an application of (D3.2).

Symmetry and transitivity are, as usual, mediate inference rules, i.e., the inference rules

$$\frac{a = b : \mathrm{el}(A)}{b = a : \mathrm{el}(A)} \tag{M4.3}$$

and

$$\frac{a = b : \mathrm{el}(A) \quad b = c : \mathrm{el}(A)}{a = c : \mathrm{el}(A)}\,, \tag{M4.4}$$

have the usual schematic demonstrations. This completes the definition of the four noncanonical forms of assertion.

There are however two inference rules which are valid for canonical elements but which, as of yet, do not have any noncanonical counterparts. The first of these inference rules is

$$\frac{a : \mathrm{el}(A) \quad A = B : \mathrm{set}}{a : \mathrm{el}(B)}\,. \tag{R4.1}$$

This rule of set conversion is recognized in virtue of the computation rule

$$\frac{\mathrm{el}(A) : a \Rightarrow c \,\varepsilon\, \mathrm{el}(C) \quad A = B : \mathrm{set} \quad (B \Rightarrow D \,\varepsilon\, \mathrm{set})}{\mathrm{el}(B) : a \Rightarrow c \,\varepsilon\, \mathrm{el}(D)}\,. \tag{C4.1}$$

The third and parenthesized premiss of this computation rule is needed only as a presupposition of the conclusion.

Two questions have to be answered: (1) Does this computation rule preserve exactness of computation? and (2) Does it preserve finiteness of computation? It is clear that it preserves exactness since the value c is the same in both premiss and conclusion.

All inference rules with a conclusion of the form $a : \mathrm{el}(A)$ are recognized in virtue of a corresponding computation rule. Leaving the above computation rule aside, every noncanonical element has a natural habitat, a set to which it principally belongs. In introducing the above computation rule, we are in effect saying that any noncanonical element can be viewed also as an element of any set equal to its natural habitat. This said, it is clear that the above computation rule preserves finiteness of computation for any noncanonical element $a : \mathrm{el}(B)$, as we can pick the computation rule to apply according to how we gained knowledge of $a : \mathrm{el}(B)$, e.g., if by set conversion, we use the computation rule presently under consideration. It follows from the premiss $a : \mathrm{el}(A)$ that the computation $\mathrm{el}(A) : a \Rightarrow c \,\varepsilon\, \mathrm{el}(C)$ is finite, and the computation of $a : \mathrm{el}(B)$ is only one step longer. One might object that by repeated application of the special case

$$\frac{\mathrm{el}(A) : a \Rightarrow c \,\varepsilon\, \mathrm{el}(C) \quad A = A : \mathrm{set}}{\mathrm{el}(A) : a \Rightarrow c \,\varepsilon\, \mathrm{el}(C)}$$

of (C4.1) we can get arbitrarily long computations; but, by the above way of picking which computation rule to apply, this special case is only used when we have used the inference rule

$$\frac{a : \mathrm{el}(A) \quad A = A : \mathrm{set}}{a : \mathrm{el}(A)}$$

in the demonstration of $a : \mathrm{el}(A)$; and even if we insist on taking such detours in our demonstrations we can still do it only a finite number of times.

The second rule of set conversion, i.e.,

$$\frac{a = b : \mathrm{el}(A) \quad A = B : \mathrm{set}}{a = b : \mathrm{el}(B)}, \tag{J4.5}$$

can now also be justified.

Justification of (J4.5). From the presuppositions of the premisses we get canonical sets and elements corresponding to all terms: $\mathrm{el}(A) : a \Rightarrow a_0 \ \varepsilon \ \mathrm{el}(A_0)$, $\mathrm{el}(A) : b \Rightarrow b_0 \ \varepsilon \ \mathrm{el}(A_0)$, and $B \Rightarrow B_0 \ \varepsilon$ set. By (C4.1), we also have $\mathrm{el}(B) : a \Rightarrow a_0 \ \varepsilon \ \mathrm{el}(B_0)$ and $\mathrm{el}(B) : b \Rightarrow b_0 \ \varepsilon \ \mathrm{el}(B_0)$. To justify that $a = b : \mathrm{el}(B)$, we need that $a_0 = b_0 \ \varepsilon \ \mathrm{el}(B_0)$. By (D4.1), we get $A_0 = B_0 \ \varepsilon$ set, and the desired conclusion follows from an application of (D3.4). This completes the justification.

To complete the generalization of the inference rules of Ch. III from the canonical to the noncanonical, I also have to give noncanonical counterparts of the inference rules of Section 5 of that chapter. Since this generalization is entirely systematic, I only show in detail how it works for the numbers. First we have the assertion

$$\mathrm{N} : \mathrm{set}. \tag{R4.2}$$

This assertion is recognized in virtue of the computation rule for the term N, viz.,

$$\mathrm{N} \Rightarrow \mathrm{N} \ \varepsilon \ \mathrm{set}. \tag{C4.2}$$

Even if we use the same name for the noncanonical and the canonical set of numbers, it has different meanings on the two sides of the arrow. The canonical set of numbers is defined by its introduction rules whereas the noncanonical set of numbers is defined by the above computation rule.[44] The same applies to the constant zero, introduced by

$$0 : \mathrm{el}(\mathrm{N}), \tag{R4.3}$$

and computed by

$$\mathrm{el}(\mathrm{N}) : 0 \Rightarrow 0 \ \varepsilon \ \mathrm{el}(\mathrm{N}). \tag{C4.3}$$

Note that this computation rule presupposes that $\mathrm{N} \Rightarrow \mathrm{N} \ \varepsilon$ set, which was laid down above. Next, for the successor, we want the inference rule

$$\frac{n : \mathrm{el}(\mathrm{N})}{\mathrm{s}(n) : \mathrm{el}(\mathrm{N})}, \tag{R4.4}$$

where the subject is computed by the rule

$$\frac{\mathrm{el}(\mathrm{N}) : n \Rightarrow m \ \varepsilon \ \mathrm{el}(\mathrm{N})}{\mathrm{el}(\mathrm{N}) : \mathrm{s}(n) \Rightarrow \mathrm{s}(m) \ \varepsilon \ \mathrm{el}(\mathrm{N})}, \tag{C4.4}$$

[44] Cf. Husserl, *Log. Unt. II*, Pt. 2, Inv. 6, § 18.

i.e., eagerly. Finally we need the equality rule

$$\frac{n = m : \mathrm{el}(\mathrm{N})}{\mathrm{s}(n) = \mathrm{s}(m) : \mathrm{el}(\mathrm{N})} \,. \tag{J4.6}$$

Justification. Let the premiss $n = m : \mathrm{el}(\mathrm{N})$ be given. The presuppositions mean that $\mathrm{el}(\mathrm{N}) : n \Rightarrow n_0 \ \varepsilon\ \mathrm{el}(\mathrm{N})$ and $\mathrm{el}(\mathrm{N}) : m \Rightarrow m_0 \ \varepsilon\ \mathrm{el}(\mathrm{N})$. By (C4.4), we have $\mathrm{el}(\mathrm{N}) : \mathrm{s}(n) \Rightarrow \mathrm{s}(n_0) \ \varepsilon\ \mathrm{el}(\mathrm{N})$ and $\mathrm{el}(\mathrm{N}) : \mathrm{s}(m) \Rightarrow \mathrm{s}(m_0) \ \varepsilon\ \mathrm{el}(\mathrm{N})$. The conclusion of the inference rule we are justifying means that $\mathrm{s}(n_0) = \mathrm{s}(m_0) \ \varepsilon\ \mathrm{el}(\mathrm{N})$. We know that $n = m : \mathrm{el}(\mathrm{N})$ wherefore, by (D4.2), we also know that $n_0 = m_0 \ \varepsilon\ \mathrm{el}(\mathrm{N})$. The desired conclusion now follows from (D3.11).

Similar generalizations from the canonical to the noncanonical could be given for the other sets introduced in Chapter III, Section 5, namely, B, 1, $A \times B$, $A + B$, and $\mathrm{L}(A, m)$, but, strictly speaking, this is not necessary since these sets and their elements are introduced for hypothetical assertions in Chapter V, Section 4, and the noncanonical sets and elements can be viewed as sets and elements under zero assumptions, i.e., in the empty context.

§ 5. Nominal definitions

A suitable next step is to consider nominal, or abbreviatory, definitions of noncanonical sets and elements. A nominal definition is a definition where a word is defined to have the same referent as another word.[45] In intuitionistic type theory, we can have nominal definitions of both sets and elements. If A : set and the expression D is not previously defined as a noncanonical set, then we can make the definition

$$D \stackrel{\mathrm{def}}{=} A : \mathrm{set}. \tag{Df}$$

The effect of such a definition is that D is a noncanonical set, the value of which is the value of A. In particular, D and A are equal sets. If we contemplate this for a moment, it becomes clear that such a nominal definition can be divided into three steps. The first step is to introduce the expression to define

$$D : \mathrm{set}, \tag{R}$$

and the second step is to give it the computation rule,

$$\frac{A \Rightarrow B \ \varepsilon\ \mathrm{set}}{D \Rightarrow B \ \varepsilon\ \mathrm{set}} \,. \tag{C}$$

These two steps have to be understood together, just as the definition of a canonical set has to be understood together with the definition of its elements. Here D : set is recognized in virtue of the computation rule. Finally, the assertion

$$D = A : \mathrm{set} \tag{J}$$

has a trivial justification.

The nominal definition of elements follows the same pattern. If $a : \mathrm{el}(A)$ and the expression d is not previously defined as a noncanonical element of the set A, or of any set equal to A, then

$$d \stackrel{\mathrm{def}}{=} a : \mathrm{el}(A) \tag{Df}$$

[45] E.g., Gredt, *Elem. Phil.* n. 33.

is used as abbreviation for introducing the expression d,

$$d : \mathrm{el}(A), \tag{R}$$

giving it a computation rule,

$$\frac{\mathrm{el}(A) : a \Rightarrow b \; \varepsilon \; \mathrm{el}(B)}{\mathrm{el}(A) : d \Rightarrow b \; \varepsilon \; \mathrm{el}(B)}\,, \tag{C}$$

and justifying that it is equal to a,

$$d = a : \mathrm{el}(A). \tag{J}$$

The condition that the expression d must not be previously defined as a noncanonical element of the set A, *or of any set equal to* A, is explained by our next topic.

Now that I have explained the meanings of equality between noncanonical sets and elements, and how nominal definitions are to be interpreted, this can be connected to the notion of definitional equality in the descriptive definition of the notion of set on p. 59. Equality between noncanonical elements respects nominal definitions, it is reflexive, and it is cancellable. There remains the criterion that two elements are equal if they have the same form and their parts are equal. This principle is established on a form-by-form basis; the case of the successor form was given in the above justification. The principle cannot even be formulated as an inference rule, because to say that something is a form is not itself a type-theoretic assertion.[46] Nevertheless, by looking through the different forms of expression introduced in this book, it becomes clear that the principle is valid, and I will call the equality between noncanonical sets and elements definitional in the sense defined by the four criteria. Using the word Thesis in the same sense as on p. 35, this can be formulated as follows:

Thesis 2. *If two canonical or noncanonical terms a and b of a certain logical category C are equal in the sense that $a = b \; \varepsilon \; C$ or $a = b : C$, then they are definitionally equal in the sense defined by the four criteria laid down on p. 32 and p. 59.*

§ 6. Functions as objects

Euler's notation for function application works well in the two cases considered above, i.e., when $f(x)$ is a function of variables or when f is a functional form. The next step in the evolution of the notion of function is that of a *function object.*[47] Since a function object is an object of a certain logical category, the function can no longer be viewed as the *form* of the application. I will write

---

[46] In fact, it cannot be a type-theoretic assertion since the subject and predicate of a predication are both expressions, and a form is not an expression by itself (unless of arity zero).

[47] This notion of function is familiar from functional programming (cf. Backus, 'Function Level Programs as Mathematical Objects').

$\text{app}(f, a)$ for the application of the function object f to the argument a, and $\text{app}[f, a]$ in the case when f and a are canonical objects.[48]

Let A and B be sets; that f is a function from A to B means, *prima facie*, that if a is a canonical element of A then $\text{app}[f, a]$ is a computational expression for an element of B, i.e., $\text{app}[f, a]$ is a noncanonical element of B. In particular, A has to be a canonical set and B a noncanonical set for this to make sense. Of course, the functions from A to B ought to form a set. That is, the four steps required for the inference rule

$$\frac{A \ \varepsilon \ \text{set} \quad B : \text{set}}{A \to B \ \varepsilon \ \text{set}} \tag{R4.5}$$

to be recognized as valid have to be performed.[49] First, that f is a canonical function from A to B means that if a is a canonical element of A then $\text{app}[f, a]$ is a noncanonical element of B, and that if a and b are equal canonical elements of A then $\text{app}[f, a]$ and $\text{app}[f, b]$ are equal noncanonical elements of B, i.e., the two inference rules

$$\frac{f \ \varepsilon \ \text{el}(A \to B) \quad a \ \varepsilon \ \text{el}(A)}{\text{app}[f, a] : \text{el}(B)} \tag{D4.3}$$

and

$$\frac{f \ \varepsilon \ \text{el}(A \to B) \quad a = b \ \varepsilon \ \text{el}(A)}{\text{app}[f, a] = \text{app}[f, b] : \text{el}(B)} \tag{D4.4}$$

are meaning determining for the form of assertion $f \ \varepsilon \ \text{el}(A \to B)$.

Next, that f and g are equal canonical functions from A to B means that if a is a canonical element of A then $\text{app}[f, a]$ and $\text{app}[g, a]$ are equal noncanonical elements of B, i.e., the inference rule

$$\frac{f = g \ \varepsilon \ \text{el}(A \to B) \quad a \ \varepsilon \ \text{el}(A)}{\text{app}[f, a] = \text{app}[g, a] : \text{el}(B)} \tag{D4.5}$$

is meaning determining for the form of assertion $f = g \ \varepsilon \ \text{el}(A \to B)$.

To complete the definition of the set $A \to B$, it has to be verified that this notion of equality between functions is reflexive and cancellable. That it is reflexive means only that $\text{app}[f, a] = \text{app}[f, a] : \text{el}(B)$ whenever a is a canonical element of A, and this follows from (D4.3) and (J4.3). To show that it is cancellable, take three elements f, g, and h, of the set $A \to B$, where $f = h : \text{el}(A \to B)$ and $g = h : \text{el}(A \to B)$. It has to be shown that $\text{app}[f, a] = \text{app}[g, a] : \text{el}(B)$ whenever a is a canonical element of A. This is demonstrated by using the definition of equality between functions, (D4.5), and that equality between elements of B is cancellable, (J4.4). Inference rule (R4.5) can now be recognized as valid.

The set $A \to B$ is unlike all of our previous examples of sets in that it has an instrumental definition,[50], i.e., it is a coinductive set; but, still, it is completely on a par with the previous examples *as a set*.

---

[48]Several different notations are used in the literature. For example: LISP $(f\,a)$, SML $f\,a$, ISWIM $f(a)$, and FP $f : a$ (cf. Steele, *Common Lisp the Language*; Milner, Tofte and Harper, *The definition of Standard ML*; Landin, 'The next 700 programming languages'; Backus, 'Can programming be liberated from the von Neumann style? '). I have chosen to make the form explicit using a notation derived from Martin-Löf, *Intuitionistic Type Theory*, p. 28.

[49]A variant notation for the same set is the power notation B^A in set theory.

[50]Instrumental definitions are explained on p. 50.

As for any set forming operation, it also has to be established that complex set of this form are equal if their parts are equal, i.e., the inference rule

$$\frac{A = C\ \varepsilon\ \text{set} \quad B = D : \text{set}}{A \to B = C \to D\ \varepsilon\ \text{set}} \tag{J4.7}$$

has to be justified.

Justification. Let the premisses $A = C\ \varepsilon$ set and $B = D$: set be given. By Definition 5 on p. 66, we have to establish four things, but, because of symmetry, i.e., (M3.3) and (M4.1), we need to consider only the first two. First, let $f\ \varepsilon\ \text{el}(A \to B)$ be given. We need to establish that $f\ \varepsilon\ \text{el}(C \to D)$, i.e., that $\text{app}[f, c] : \text{el}(D)$ whenever $c\ \varepsilon\ \text{el}(C)$ and that $\text{app}[f, c] = \text{app}[f, d] : \text{el}(D)$ whenever $c = d\ \varepsilon\ \text{el}(C)$. We argue thus: $c\ \varepsilon\ \text{el}(C)$, therefore, by (D3.5), $c\ \varepsilon\ \text{el}(A)$, therefore, by (D4.3), $\text{app}[f, c] : \text{el}(B)$, therefore, by (R4.1), $\text{app}[f, c] : \text{el}(D)$, as required (the premisses $A = C\ \varepsilon$ set, $f\ \varepsilon\ \text{el}(A \to B)$, and $B = D$: set, needed in these three steps, are already known). Similarly, we argue: $c = d\ \varepsilon\ \text{el}(C)$, therefore, by (D3.6), $c = d\ \varepsilon\ \text{el}(A)$, therefore, by (D4.4), $\text{app}[f, c] = \text{app}[f, d] : \text{el}(B)$, therefore, by (J4.5), $\text{app}[f, c] = \text{app}[f, d] : \text{el}(D)$. Next, let $f = g\ \varepsilon\ \text{el}(A \to B)$ be given. We need to establish that $f = g\ \varepsilon\ \text{el}(C \to D)$, i.e., that $\text{app}[f, c] = \text{app}[g, c] : \text{el}(C \to D)$ whenever $c\ \varepsilon\ \text{el}(C)$. We argue thus: $c\ \varepsilon\ \text{el}(C)$, therefore, by (D3.5), $c\ \varepsilon\ \text{el}(A)$, therefore, by (D4.5), $\text{app}[f, c] = \text{app}[g, c] : \text{el}(B)$, therefore, by (J4.5), $\text{app}[f, c] = \text{app}[g, c] : \text{el}(D)$, as required. This completes the justification.

I have demanded, as a general principle, that any form must respect equality. What about the form app with canonical parts? The inference rule

$$\frac{f = g\ \varepsilon\ \text{el}(A \to B) \quad a = b\ \varepsilon\ \text{el}(A)}{\text{app}[f, a] = \text{app}[g, b] : \text{el}(B)} \tag{M4.5}$$

can in fact be demonstrated. One way to do it is by using $\text{app}[g, a]$ as middle term:

$$\frac{\dfrac{f = g\ \varepsilon\ \text{el}(A \to B) \quad a\ \varepsilon\ \text{el}(A)}{\text{app}[f, a] = \text{app}[g, a] : \text{el}(B)} \quad \dfrac{g\ \varepsilon\ \text{el}(A \to B) \quad a = b\ \varepsilon\ \text{el}(A)}{\text{app}[g, a] = \text{app}[g, b] : \text{el}(B)}}{\text{app}[f, a] = \text{app}[g, b] : \text{el}(B)} .$$

I have only defined the application of a canonical function f to a canonical argument a, but the definition can be extended to the case when both f and a are noncanonical elements. First, the inference rule

$$\frac{A : \text{set} \quad B : \text{set}}{A \to B : \text{set}} , \tag{R4.6}$$

has the computation rule

$$\frac{A \Rightarrow C\ \varepsilon\ \text{set}}{A \to B \Rightarrow C \to B\ \varepsilon\ \text{set}} . \tag{C4.5}$$

Note that B is not evaluated; the form $A \to B$ is eager in A but lazy in B. The noncanonical application rule is now formulated as

$$\frac{f : \text{el}(A \to B) \quad a : \text{el}(A)}{\text{app}(f, a) : \text{el}(B)} , \tag{R4.7}$$

and recognized as valid in virtue of the computation rule

$$\frac{\begin{array}{l} \text{el}(A) : a \Rightarrow c\ \varepsilon\ \text{el}(C) \\ \text{el}(A \to B) : f \Rightarrow g\ \varepsilon\ \text{el}(C \to B) \\ \text{el}(B) : \text{app}[g, c] \Rightarrow d\ \varepsilon\ \text{el}(D) \end{array}}{\text{el}(B) : \text{app}(f, a) \Rightarrow d\ \varepsilon\ \text{el}(D)} . \tag{C4.6}$$

The reader is advised to verify that this computation rule is well-formed.

Note that the sets A and B are not made explicit in the notation $\mathrm{app}(f, a)$, and that this can lead to ambiguities when f and a are polymorphic expressions; for example, if the polymorphic expression f can be viewed as an element of both the set $A_1 \rightarrow B$ and the set $A_2 \rightarrow B$ and the polymorphic expression a can be viewed as an element of both A_1 and A_2, then $\mathrm{app}(f, a)$ is ambiguous.[51] On the other hand, it is too hampering to make A and B explicit for each application: thus, I will keep the possibly ambiguous notation, since in all practical cases it will be clear which function it is that is to be applied to what argument; put differently, the possibly ambiguous notation may not be used when it really is ambiguous.[52]

To complete the extension of functions from the canonical to the noncanonical, it has to be established that the noncanonical forms respect equality, i.e., that

$$\frac{A = C : \mathrm{set} \quad B = D : \mathrm{set}}{A \rightarrow B = C \rightarrow D : \mathrm{set}}, \tag{J4.8}$$

and that

$$\frac{f = g : \mathrm{el}(A \rightarrow B) \quad a = b : \mathrm{el}(A)}{\mathrm{app}(f, a) = \mathrm{app}(g, b) : \mathrm{el}(B)}. \tag{J4.9}$$

Since the justifications are not completely analogous to any previous justification, it is worth looking at them in some detail.

Justification of (J4.8). Let the premisses $A = C : \mathrm{set}$ and $B = D : \mathrm{set}$ be given. The presuppositions of the premisses mean that $A \Rightarrow A_0 \;\varepsilon\; \mathrm{set}$ and $C \Rightarrow C_0 \;\varepsilon\; \mathrm{set}$. Since, by (C4.5), $A \rightarrow B \Rightarrow A_0 \rightarrow B \;\varepsilon\; \mathrm{set}$ and $C \rightarrow D \Rightarrow C_0 \rightarrow D \;\varepsilon\; \mathrm{set}$, the conclusion means that $A_0 \rightarrow B = C_0 \rightarrow D \;\varepsilon\; \mathrm{set}$. By (D4.1), $A_0 = C_0 \;\varepsilon\; \mathrm{set}$, and the desired conclusion follows from (J4.7). This completes the justification.

Justification of (J4.9). Let the premisses $f = g : \mathrm{el}(A \rightarrow B)$ and $a = b : \mathrm{el}(A)$ be given. Their presuppositions mean that $\mathrm{el}(A) : a \Rightarrow a_0 \;\varepsilon\; \mathrm{el}(A_0)$, $\mathrm{el}(A) : b \Rightarrow b_0 \;\varepsilon\; \mathrm{el}(A_0)$, $\mathrm{el}(A \rightarrow B) : f \Rightarrow f_0 \;\varepsilon\; \mathrm{el}(A_0 \rightarrow B)$, and $\mathrm{el}(A \rightarrow B) : g \Rightarrow g_0 \;\varepsilon\; \mathrm{el}(A_0 \rightarrow B)$. From this we get $f_0[a_0] : \mathrm{el}(B)$ and $g_0[b_0] : \mathrm{el}(B)$, by (D4.3). This means that $\mathrm{el}(B) : f_0[a_0] \Rightarrow d \;\varepsilon\; \mathrm{el}(D)$ and $\mathrm{el}(B) : g_0[b_0] \Rightarrow e \;\varepsilon\; \mathrm{el}(D)$, for some canonical elements d and e of the set D which is the value of B. By (C4.6), we also have $\mathrm{el}(B) : \mathrm{app}(f, a) \Rightarrow d \;\varepsilon\; \mathrm{el}(D)$ and $\mathrm{el}(B) : \mathrm{app}(g, b) \Rightarrow e \;\varepsilon\; \mathrm{el}(D)$. It remains to show that $d = e \;\varepsilon\; \mathrm{el}(D)$. That $a_0 = b_0 : \mathrm{el}(A_0)$ and $f_0 = g_0 : \mathrm{el}(A_0 \rightarrow B)$ both follow from (D4.2). By (M4.5), $f_0[a_0] = g_0[b_0] : \mathrm{el}(B)$, and the desired conclusion follows from a final application of (D4.2). This completes the justification.

This completes the definition of the set of functions from A to B.

One could now continue by introducing binary and ternary functions but, strictly speaking, this is not needed since a binary function can be viewed either as an element of the set $(A \times B) \rightarrow C$ or as an element of the set $A \rightarrow (B \rightarrow C)$. In fact, it can be demonstrated, in the language of intuitionistic type theory, that these two sets are isomorphic in the usual sense. In the former case, the application of a function f to two arguments a and b is written $\mathrm{app}(f, (a, b))$;

---

[51]Cf. p. 92 and p. 101.

[52]If ambiguous expressions are to be allowed, the inference rules of intuitionistic type theory cannot be understood completely formally, as demonstrated by the paradoxical result gained by Salvesen, *Polymorphism and monomorphism in Martin-Löf's Type Theory*, pp. 20–30. Note that it is *possible* to allow for ambiguous expressions without ending up in these paradoxes (ibid., Theorems 4–7) if different occurences of an ambiguous expression are taken as standing for different terms (cf. p. 14 of this book).

in the latter case, the application of a function g to two arguments is written $\mathrm{app}(\mathrm{app}(g, a), b)$. The first way of writing function application is perhaps more natural, but sometimes the latter form is preferable. Sometimes g is called the Curried form of f, and the translation which takes f to g is called Currying.[53]

§ 7. Families of sets

I will now consider a special kind of functions, namely, the set valued functions or families of sets.

Definition 10. *That F is a set valued function on the canonical set A, abbreviated $F \mathrel{\varepsilon} \mathrm{fam}(A)$, means that if a is a canonical element of A then $\mathrm{app}[F, a]$ is a noncanonical set, and that equal canonical elements of A give equal noncanonical sets.*

This definition introduces a new form of assertion, $F \mathrel{\varepsilon} \mathrm{fam}(A)$, with the presupposition that A is a canonical set. The inference rules,

$$\frac{F \mathrel{\varepsilon} \mathrm{fam}(A) \quad a \mathrel{\varepsilon} \mathrm{el}(A)}{\mathrm{app}[F, a] : \mathrm{set}} \tag{D4.6}$$

and

$$\frac{F \mathrel{\varepsilon} \mathrm{fam}(A) \quad a = b \mathrel{\varepsilon} \mathrm{el}(A)}{\mathrm{app}[F, a] = \mathrm{app}[F, b] : \mathrm{set}} \tag{D4.7}$$

are immediate from this definition and they completely characterize what it means to be a set valued function on the canonical set A.

It must also be defined what it means for two set valued functions to be equal.

Definition 11. *That F and G are equal set valued functions on the canonical set A, abbreviated $F = G \mathrel{\varepsilon} \mathrm{fam}(A)$, means that if a is a canonical element of A then $\mathrm{app}[F, a]$ and $\mathrm{app}[G, a]$ are equal noncanonical sets.*

This definition introduces the new form of assertion $F = G \mathrel{\varepsilon} \mathrm{fam}(A)$, with the presuppositions that F and G are set valued functions on the canonical set A. The following inference rule is immediate from the definition and characterizes what it means for two set valued functions to be equal

$$\frac{F = G \mathrel{\varepsilon} \mathrm{fam}(A) \quad a \mathrel{\varepsilon} \mathrm{el}(A)}{\mathrm{app}[F, a] = \mathrm{app}[G, a] : \mathrm{set}} \,. \tag{D4.8}$$

The equality relation so defined is reflexive

$$\frac{F \mathrel{\varepsilon} \mathrm{fam}(A)}{F = F \mathrel{\varepsilon} \mathrm{fam}(A)} \tag{J4.10}$$

and cancellable

$$\frac{F = H \mathrel{\varepsilon} \mathrm{fam}(A) \quad G = H \mathrel{\varepsilon} \mathrm{fam}(A)}{F = G \mathrel{\varepsilon} \mathrm{fam}(A)} \,. \tag{J4.11}$$

[53]The term Currying is actually a misnomer, apparently due to Strachey. Gödel, 'Über eine bisher noch nicht benütze Erweiterung des finiten Standpunktes', fn. 8, refers to Church, and Church, 'A Formulation of the Simple Theory of Types', p. 57, refers to Schönfinkel, but the technique was actually used already by Frege in *Grundgesetze der Arithmetik I*, §§ 35-37. As Fregeing does not sound very nice, I will stick to the well-established term Currying.

Justification of (J4.10). Let the premiss $F \,\varepsilon\, \text{fam}(A)$ be given. The conclusion means that $\text{app}[F,a] = \text{app}[F,a] : \text{set}$ whenever $a \,\varepsilon\, \text{el}(A)$. But, granted that $a \,\varepsilon\, \text{el}(A)$, we have $\text{app}[F,a] : \text{set}$, by (D4.6), and $\text{app}[F,a] = \text{app}[F,a] : \text{set}$, by (J4.1), as required.

Justification of (J4.11). Let the premisses $F = H \,\varepsilon\, \text{fam}(A)$ and $G = H \,\varepsilon\, \text{fam}(A)$ be given. The conclusion means that $\text{app}[F,a] = \text{app}[G,a] : \text{set}$ whenever $a \,\varepsilon\, \text{el}(A)$. But, granted that $a \,\varepsilon\, \text{el}(A)$, we have $\text{app}[F,a] = \text{app}[H,a] : \text{set}$ and $\text{app}[G,a] = \text{app}[H,a] : \text{set}$, by (D4.8), and it follows from (J4.2) that $\text{app}[F,a] = \text{app}[G,a] : \text{set}$.

As usual, if an equality relation is reflexive and cancellable, it is also symmetric

$$\frac{F = G \,\varepsilon\, \text{fam}(A)}{G = F \,\varepsilon\, \text{fam}(A)} \tag{M4.6}$$

and transitive

$$\frac{F = G \,\varepsilon\, \text{fam}(A) \quad G = H \,\varepsilon\, \text{fam}(A)}{F = H \,\varepsilon\, \text{fam}(A)}\,. \tag{M4.7}$$

The schematic demonstrations are the same as those on p. 66, *mutatis mutandis.*

Next, consider the following two inference rules

$$\frac{A = B \,\varepsilon\, \text{set} \quad F \,\varepsilon\, \text{fam}(A)}{F \,\varepsilon\, \text{fam}(B)} \tag{J4.12}$$

and

$$\frac{A = B \,\varepsilon\, \text{set} \quad F = G \,\varepsilon\, \text{fam}(A)}{F = G \,\varepsilon\, \text{fam}(B)}\,. \tag{J4.13}$$

Justification of (J4.12). Let the premisses $A = B \,\varepsilon\, \text{set}$ and $F \,\varepsilon\, \text{fam}(A)$ be given. The conclusion means that $\text{app}[F,a] : \text{set}$ whenever $a \,\varepsilon\, \text{el}(A)$ and that $\text{app}[F,a] = \text{app}[F,b] : \text{set}$ whenever $a = b \,\varepsilon\, \text{el}(A)$. The former follows from (D3.5) and (D4.6). The latter follows from (D3.6) and (D4.7). This completes the justification.

Justification of (J4.13). Let the premisses $A = B \,\varepsilon\, \text{set}$ and $F = G \,\varepsilon\, \text{fam}(A)$ be given. The conclusion means that $\text{app}[F,a] = \text{app}[G,a] : \text{set}$ whenever $a \,\varepsilon\, \text{el}(B)$, which follows from (D3.5) and (D4.8).

Since equality between canonical sets is symmetric, the two inference rules justified above mean that if $A = B \,\varepsilon\, \text{set}$ then the two logical categories $\text{fam}(A)$ and $\text{fam}(B)$ are logically interchangeable.

The mediate inference rule

$$\frac{F = G \,\varepsilon\, \text{fam}(A) \quad a = b \,\varepsilon\, \text{el}(A)}{\text{app}[F,a] = \text{app}[G,b] : \text{set}}\,, \tag{M4.8}$$

has the schematic demonstration

$$\frac{\dfrac{F \,\varepsilon\, \text{fam}(A) \quad a = b \,\varepsilon\, \text{el}(A)}{\text{app}[F,a] = \text{app}[F,b] : \text{set}} \quad \dfrac{F = G \,\varepsilon\, \text{fam}(A) \quad b \,\varepsilon\, \text{el}(A)}{\text{app}[F,b] = \text{app}[G,b] : \text{set}}}{\text{app}[F,a] = \text{app}[G,b] : \text{set}}\,.$$

A set valued function on a canonical set A is also called a family of sets over A. This terminology is based on the idea that a set valued function, F say, defined on a canonical set

$$A = \{a_1, a_2, a_2, \ldots\}$$

defines the "family" of sets

$$\{\text{app}[F,a_1], \text{app}[F,a_2], \text{app}[F,a_3], \ldots\}$$

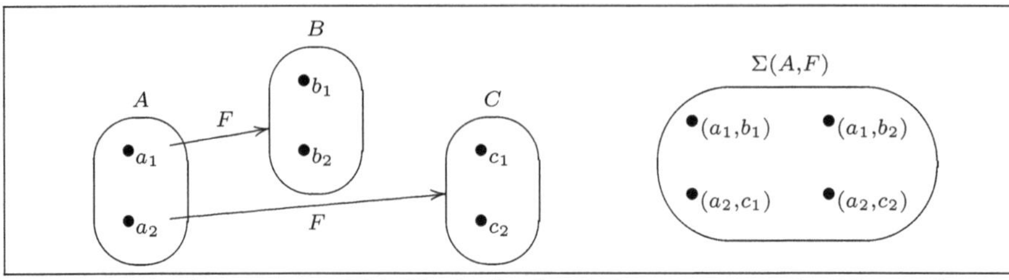

Figure 7. The disjoint union $\Sigma(A,F)$ of a family F of sets over a set A with two elements a_1 and a_2 which takes a_1 to the set $B = \{b_1, b_2\}$ and a_2 to the set $C = \{c_1, c_2\}$.

Instead of family one could call it class or collection of sets, but I will still consider two families, classes, or collections of sets equal only if they are intensionally equal as set valued functions.

An example of a family of sets is the family F which takes a number n to the set $\mathrm{L}(\mathrm{N}, n)$, i.e.,

$$\mathrm{app}[\mathrm{F}, n] \Rightarrow \mathrm{L}(\mathrm{N}, n) \ \varepsilon \ \mathrm{set},$$

and $\mathrm{F} \ \varepsilon \ \mathrm{fam}(\mathrm{N})$. Further examples will be given later.

There are two natural set forming operations on a family of sets: the sum and the product.

First, I consider the sum. If A and F are as above, I like to make rigorous the notion of a possibly infinite sum

$$\mathrm{app}[F, a_1] + \mathrm{app}[F, a_2] + \mathrm{app}[F, a_3] + \cdots$$

of sets. Contemplating the matter, it becomes clear that the outstanding candidate for a canonical element of this sum is a pair of two elements, the first component of which is a canonical element a of A, telling us which set $\mathrm{app}[F, a]$ the second component (also canonical) of the pair belongs to. If the sum of a family of sets F over the canonical set A is denoted by $\Sigma(A, F)$,[54] we get the inference rule

$$\frac{A \ \varepsilon \ \mathrm{set} \quad F \ \varepsilon \ \mathrm{fam}(A)}{\Sigma(A, F) \ \varepsilon \ \mathrm{set}} \ . \tag{R4.8}$$

The above explanation of what counts as an element of the sum is translated into the inference rule

$$\frac{a \ \varepsilon \ \mathrm{el}(A) \quad (F \ \varepsilon \ \mathrm{fam}(A)) \quad \mathrm{app}[F, a] \Rightarrow B \ \varepsilon \ \mathrm{set} \quad b \ \varepsilon \ \mathrm{el}(B)}{(a\,, b) \ \varepsilon \ \mathrm{el}(\Sigma(A, F))} \ . \tag{D4.9}$$

Two such elements are equal if their parts are equal, i.e.,

$$\frac{a = c \ \varepsilon \ \mathrm{el}(A) \quad (F \ \varepsilon \ \mathrm{fam}(A)) \quad \mathrm{app}[F, a] \Rightarrow B \ \varepsilon \ \mathrm{set} \quad b = d : \mathrm{el}(B)}{(a\,, b) = (c\,, d) \ \varepsilon \ \mathrm{el}(\Sigma(A, F))} \ . \tag{D4.10}$$

As before, premisses are put within parentheses if they are needed only as presuppositions of the conclusion.

This definition is well-formed because, by (D4.7), $\mathrm{app}[F, a] = \mathrm{app}[F, c] : \mathrm{set}$ and, if $\mathrm{app}[F, c] \Rightarrow C \ \varepsilon \ \mathrm{set}$, $B = C \ \varepsilon \ \mathrm{set}$, by (D4.1), whence $d \ \varepsilon \ \mathrm{el}(C)$, by (D3.3), as required by the second presupposition of the conclusion.

---

[54]The symbol Σ for sums of numbers was introduced by Euler, *Institutiones Calculi Differentialis*, p. 27.

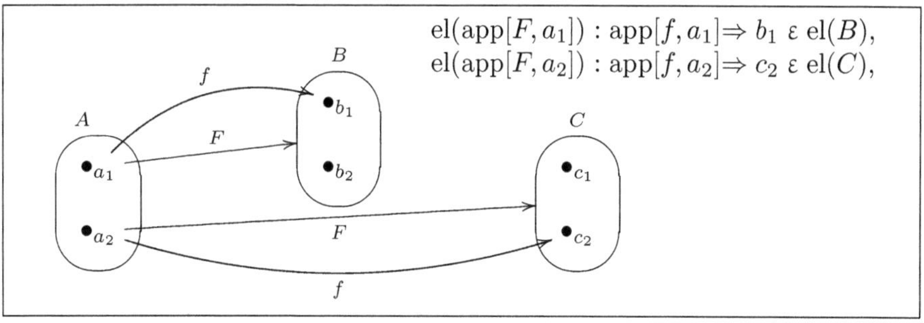

Figure 8. Diagram exemplifying a dependently typed function f on a set A with two elements a_1 and a_2 ; f takes a_1 to b_1 and a_2 to c_2.

In view of (D3.1), it is clear that this equality relation is reflexive. It is left as an easy exercise to make sure that it is cancellable.

As usual, it must also be verified that the form Σ respects equality, i.e., that the inference rule

$$\frac{A = B \ \varepsilon \ \mathrm{set} \quad F = G \ \varepsilon \ \mathrm{fam}(A)}{\Sigma(A, F) = \Sigma(B, G) \ \varepsilon \ \mathrm{set}} \tag{J4.14}$$

is valid. Observe that the inference rule is well-formed because of (J4.12). This easy but tedious justification is left to the reader.

The set $\Sigma(A, F)$ is also called the disjoint union of a family of sets, because it corresponds to a construction in extensional set theory where first the sets $\mathrm{app}[F, a]$ are made disjoint and then their extensional union is taken. However, in intuitionistic type theory, there is no extensional union of a family of sets and the construction of the sum is instead a primitive operation.

Having thus defined the sum of a family of sets, I turn to the product. Let A and F be defined as above. The notion of a possibly infinite product

$$\mathrm{app}[F, a_1] \times \mathrm{app}[F, a_2] \times \mathrm{app}[F, a_3] \times \cdots$$

of sets must now be made rigorous. Since the symbol Π is used for products of numbers, I will use it also for sets. The four steps necessary to recognize as valid the inference rule

$$\frac{A \ \varepsilon \ \mathrm{set} \quad F \ \varepsilon \ \mathrm{fam}(A)}{\Pi(A, F) \ \varepsilon \ \mathrm{set}} \ . \tag{R4.9}$$

now have to be taken. An element of the set $\Pi(A, F)$ must consist of an element of $\mathrm{app}[F, a_1]$, an element of $\mathrm{app}[F, a_2]$, an element of $\mathrm{app}[F, a_3]$, etc., but how is this expressed in the language of type theory? It is made into the general rule that an element of $\Pi(A, F)$ consists of an element of $\mathrm{app}[F, a]$ whenever a is a canonical element of A. If we use the name $\mathrm{app}[f, a]$ for this distinguished element of $\mathrm{app}[F, a]$, we arrive at the inference rule

$$\frac{f \ \varepsilon \ \mathrm{el}(\Pi(A, F)) \quad a \ \varepsilon \ \mathrm{el}(A)}{\mathrm{app}[f, a] : \mathrm{el}(\mathrm{app}[F, a])} \ , \tag{D4.11}$$

which I take as defining what it means to be an element of the set $\Pi(A, F)$, under the condition that it respects equality, i.e., under the condition

$$\frac{f \ \varepsilon\ \mathrm{el}(\Pi(A, F)) \quad a = b \ \varepsilon\ \mathrm{el}(A)}{\mathrm{app}[f, a] = \mathrm{app}[f, b] : \mathrm{el}(\mathrm{app}[F, b])} \ . \tag{D4.12}$$

Note that this inference rule is well-formed because, due to (D4.7), $\mathrm{app}[F, a] = \mathrm{app}[F, b] : \mathrm{set}$ and, due to (R4.1), $\mathrm{app}[f, a] : \mathrm{el}(\mathrm{app}[F, b])$.

These inference rules explain the need for the complicated computational form of assertion

$$\mathrm{el}(A) : a \Rightarrow b \ \varepsilon\ \mathrm{el}(B).$$

In particular, it explains why the computations $A \Rightarrow B$ and $a \Rightarrow b$ cannot be separated but have to be understood together.

Upon comparing this definition to the definition of the set $A \to B$, it becomes clear that an element of $\Pi(A, F)$ is in fact a kind of function where the type of the value depends on the argument. This dependency is the origin of the word dependent in dependent type theory.[55]

Two such functions f and g are considered equal if $\mathrm{app}[f, a]$ and $\mathrm{app}[g, a]$ are equal elements of the set $\mathrm{app}[F, a]$ whenever a is a canonical element of the set A, i.e.,

$$\frac{f = g \ \varepsilon\ \mathrm{el}(\Pi(A, F)) \quad a \ \varepsilon\ \mathrm{el}(A)}{\mathrm{app}[f, a] = \mathrm{app}[g, a] : \mathrm{el}(\mathrm{app}[F, a])} \ . \tag{D4.13}$$

That this equality relation is follows from (D4.11) and (J4.3) and that it is cancellable follows from (D4.13) and (J4.4). As usual, it also has to be verified that two complex terms of Π form are equal if their parts are equal, i.e., that

$$\frac{A = B \ \varepsilon\ \mathrm{set} \quad F = G \ \varepsilon\ \mathrm{fam}(B)}{\Pi(A, F) = \Pi(B, G) \ \varepsilon\ \mathrm{set}} \ . \tag{J4.15}$$

The justification is similar to that of (J4.7).

Finally the inference rule

$$\frac{f = g \ \varepsilon\ \mathrm{el}(\Pi(A, F)) \quad a = b \ \varepsilon\ \mathrm{el}(A)}{\mathrm{app}[f, a] = \mathrm{app}[g, b] : \mathrm{el}(\mathrm{app}[F, b])} \ . \tag{M4.9}$$

has a schematic demonstration similar to that of (M4.5).

[55]Special cases of type dependency go a long way back, but it seems as if the first to recognize its importance as an independent concept was de Bruijn (cf. e.g. the 1968 report *Automath, a language for mathematics*), and, independently, Martin-Löf (cf. e.g. the 1973 article 'An intuitionistic theory of types'). Phenomena which resemble type dependency occur naturally in language (cf. Ranta, *Type-Theoretical Grammar*).

CHAPTER V

Assumption and Substitution

THE FORMS of assertion concerning canonical and noncanonical sets and elements, introduced in the previous chapters, are all *categorical* in the sense that the terms of the assertions may not depend on any assumptions. In this chapter, the type-theoretic forms of assertion will be generalized to the hypothetical case, i.e., to the case where the terms may depend on assumptions. In the first section, hypothetical assertions are related to the concept of function; the definitions of the hypothetical forms of assertions follow in the next section. In the third section I present a version of the type-theoretic substitution calculus. The fourth, fifth, and sixth sections are concerned with the generalization of the material of previous chapters to the hypothetical case. The elimination rules of intuitionistic type theory are given in the seventh section. The eighth and last section of this chapter deals with the Curry-Howard correspondence.

§ 1. The concept of function revisited

The word *function* is used in several different senses in this book: these are summarized in Table 6. The first distinction is between old-fashioned functions of variables, in the sense of Euler and Frege, where application is substitution, and functions in the modern sense of Dedekind and Riemann. For example, the expression

$$x^2 - 3x$$

is a function of the variable x whereas the function f defined by

$$\mathrm{f}(x) = x^2 - 3x$$

is a function in the modern sense of the word.

Functions in the modern sense can either be considered as a functional forms, i.e., as syncategorems or mere function symbols, or as mathematical objects in their own right. The notion of a functional form, introduced on p. 85, is a very fundamental metalinguistic notion; I find it sufficiently intelligible in itself, not to merit further elaboration. On the other hand, the notion of function object is more sophisticated: function objects always belong to some logical category; three such logical categories were introduced above, namely, $\mathrm{el}(A \to B)$, $\mathrm{fam}(A)$, and $\mathrm{el}(\Pi(A, F))$. A further distinction can be made between between different categories of function objects, viz., between categories that are *inductively* defined and those that are *coinductively* defined.[1]

[1] It is the *inductively* defined function set (Nordström, Petersson and Smith, 'Martin-Löf's Type Theory', p. 24) that corresponds to Frege's course-of-values (Wertverlauf). However,

J.G. Granström, *Treatise on Intuitionistic Type Theory*, Logic, Epistemology, and the Unity of Science 22, DOI 10.1007/978-94-007-1736-7_5,

The notion of function is divided into:	(A) The old-fashioned notion of function of variables, which can be explained in three ways:	(1) The nonrigorous and formalistic notion of function of variables, in which sense a function is an analytic expression, defined on p. 82;	
		(2) the notion of function of variables defined by informal substitution used by Martin-Löf, *Intuitionistic Type Theory*, pp. 16–20; and	
		(3) the rigorous notion of function of variables defined by substitution, defined on p. 112, sqq.; this is a clarification of the previous notions.	
	(B) Functions in the modern, or black-box, sense of which there are two kinds:	(1) The notion of *functional form* according to which the unsaturated form of an expression is called a function, defined on p. 85; and	
		(2) the notion of function as an object belonging to some logical category, which can be one of	(a) $\mathrm{el}(A \to B)$, p. 99,
			(b) $\mathrm{fam}(A)$, p. 102,
			(c) $\mathrm{el}(\Pi(A, F))$, p. 105.

Table 6. Overview of seven different notion of function considered in this book, with references to the pages on which they are defined.

While there are *two distinct notions* of function in the modern sense, the notion of function of variables is better described as *one notion* going through stages of refinement, i.e., one notion that receives more precise explanations. Euler's notion of functions of variables as analytic expressions, introduced on p. 82 and associated with the notation

$$\begin{array}{c} x : \mathrm{el}(A) \\ \vdots \\ f(x) : \mathrm{el}(B), \end{array}$$

is the first stage of refinement. The problem with this notion of function is that the mandatory inference rules

$$\frac{\begin{array}{c} (x : \mathrm{el}(A)) \\ \vdots \\ f(x) : \mathrm{el}(B) \end{array} \quad a : \mathrm{el}(A)}{f(a) : \mathrm{el}(B)}$$

and

$$\frac{\begin{array}{c} (x : \mathrm{el}(A)) \\ \vdots \\ f(x) : \mathrm{el}(B) \end{array} \quad a = b : \mathrm{el}(A)}{f(a) = f(b) : \mathrm{el}(B)} ,$$

strictly speaking, cannot be justified, i.e., they are not evident from the meanings of the terms involved. In these inference rules, I have put the premiss

the inductively defined set of functions requires the higher type structure and will play no further role in this book.

$x : \mathrm{el}(A)$ within parentheses to show that the conclusions of these inference rules no longer depend on this assumption. The expression $f(a)$ is to be understood as the expression $f(x)$ with all relevant occurrences of x replaced by a. These two substitution rules can however be *motivated* by appeal to how the term $f(x)$ is built up, or by appeal to a characteristic of the notion of definitional equality.[2]

The second stage of refinement for the notion of function of variables comes from a change of attitude towards the notation

$$\begin{array}{c} x : \mathrm{el}(A) \\ \vdots \\ f(x) : \mathrm{el}(B). \end{array}$$

As explained on p. 82, this notation stands for a formal demonstration that $f(x)$ is an element of B from the premiss $x : \mathrm{el}(A)$. The change of attitude consists in viewing

$$\begin{array}{c} (x : \mathrm{el}(A)) \\ \vdots \\ f(x) : \mathrm{el}(B) \end{array}$$

as a *form of assertion* which *means* that the two substitution rules, formulated above, are valid. To make this change of attitude more precise, a change in notation is called for. The new form of assertion

$$\begin{array}{c} (x : \mathrm{el}(A)) \\ f(x) : \mathrm{el}(B), \end{array}$$

i.e., without the four dots, is defined to mean that if a is an element of A, then $f(a)$ is an element of B, and if a and b are equal elements of A, then $f(a)$ and $f(b)$ are equal elements of B, i.e., the inference rules

$$\frac{\begin{array}{l} (x : \mathrm{el}(A)) \\ f(x) : \mathrm{el}(B) \quad a : \mathrm{el}(A) \end{array}}{f(a) : \mathrm{el}(B)}$$

and

$$\frac{\begin{array}{l} (x : \mathrm{el}(A)) \\ f(x) : \mathrm{el}(B) \quad a = b : \mathrm{el}(A) \end{array}}{f(a) = f(b) : \mathrm{el}(B)}$$

are meaning determining for their major premiss. This was the approach taken by Martin-Löf in his 1984 book.[3] The crucial difference that this change of attitude brings about is that

$$\begin{array}{c} (x : \mathrm{el}(A)) \\ f(x) : \mathrm{el}(B) \end{array}$$

is a *complete form of assertion*, whereas

$$\begin{array}{c} x : \mathrm{el}(A) \\ \vdots \\ f(x) : \mathrm{el}(B) \end{array}$$

[2]Cf. p. 32 and p. 59.

[3]*Intuitionistic Type Theory*, pp. 16–20.

is a schematic notation properly belonging to the metalanguage of intuitionistic type theory.

Another way to come to the conclusion that a new form of assertion is necessary to properly understand functions of variables is to observe that there is no such thing as an *assertion* made under an *assumption*. Recall that, when making an inference, we pass from something we *know*, the premisses, to something we *get to know*, the conclusion. The notion we attempt to capture is that $f(x)$ is an element of B *under the assumption* that x is an element of A. As mentioned in Chapter II, Section 8, a distinction must be made between the schematic letters occurring in the inference rules, which are taken to be known, and variables, which are taken to be unknown. The very notation

$$\begin{array}{c} x : \mathrm{el}(A) \\ \vdots \\ f(x) : \mathrm{el}(B) \end{array}$$

contains a confusion between the two epistemic attitudes *known* vs. *unknown*, since the x in the premiss, which should be known, is really unknown. The problem becomes apparent, again, in the substitution rules, where the assumption that x is an element of A is discharged. How a variable or schematic letter is to be taken is expressed as follows in words: "assume that x is an element of A" vs. "let an element a of A be given". In the former case, we *assume* that x is an *unknown* element of A; in the latter case, we *let* a be a *given*, or *known*, element of A.

Still, one problem remains with the form of assertion

$$\begin{array}{c} (x : \mathrm{el}(A)) \\ f(x) : \mathrm{el}(B), \end{array}$$

namely, that substitution is treated as something metamathematical, i.e., as something taking place outside the language of intuitionistic type theory. There are two problems with this approach. First, that the metamathematical treatment of substitution does not fit well with an extensible language like intuitionistic type theory; this is because inference rules cannot be justified by induction over all well-formed expressions in the absence of a fixed syntax. Next, that, in the presence of variable binding operations, substitution is not as innocuous as it seems.

"Despite the simplicity of the substitution idea, it turns out to be surprisingly complicated to give a rigorous mathematical definition of the substitution process. The problem arises from the possibility of confusion between the names used for the formal parameters of a procedure and the (possibly identical) names used in the expressions to which the procedure may be applied. Indeed, there is a long history of erroneous definitions of *substitution* in the literature of logic and programming semantics."[4]

The problem of substitution becomes visible when an open expression is substituted into an expression with bound variables; for example, if $x + z$ is

[4]Abelson and Sussman, *Structure and Interpretation of Computer Programs*, Ch. 1, fn. 15.

substituted for y in the expression

$$\int_a^b (x^2 + y^2)\,\mathrm{d}x,$$

the naïve replacement

$$\int_a^b (x^2 + (x+z)^2)\,\mathrm{d}x$$

is wrong, since the x in $x+z$ is, as it were, accidentally captured by the variable x bound by the integral. Instead one has to change the variable of integration, to w say, and then make the replacement; so, a correct result of the substitution is

$$\int_a^b (w^2 + (x+z)^2)\,\mathrm{d}w.$$

Euler's notion of function of variables, covering both that which is plugged in between the integral and the differential signs, and the complete expression for the integral, is sufficiently clear to make the *art* of substitution, and *a fortiori* the *art* of integration, work;[5] it is still taught this way in basic calculus. However, the *scientific* treatment of substitution is comparatively recent.

§ 2. Hypothetical assertions

The third stage of refinement of the notion of function of variables is to bring the *substitutions* into the language of intuitionistic type theory by means of a calculus of *explicit substitution*.[6] This solves both problems, mentioned above, with the metamathematical treatment of substitution. Explicit substitution is treated of in the present section and the next. In the form of assertion

$$\begin{array}{c}(x : \mathrm{el}(A))\\ f(x) : \mathrm{el}(B),\end{array}$$

introduced above, $f(x)$ was taken as standing for an expression into which different elements a of the set A can be substituted. If this notion of substitution is regarded as not being sufficiently clear, another explanation has to be found. That b is an element of the set B, under the assumption that x is an element of A, which I will write

$$b : \mathrm{el}(B)\ (x : \mathrm{el}(A)),$$

i.e., with the assumption to the right instead of above, is explained as follows: that $b : \mathrm{el}(B)\ (x : \mathrm{el}(A))$ means that if $a\ \varepsilon\ \mathrm{el}(A)$, then $b\,|(x \leftarrow a)$ is an element of B, and if $a = c\ \varepsilon\ \mathrm{el}(A)$, then $b\,|(x \leftarrow a)$ and $b\,|(x \leftarrow c)$ are equal elements of B. Thus, the form of assertion $b : \mathrm{el}(B)\ (x : \mathrm{el}(A))$ presupposes that A is a canonical set and that B is a noncanonical set, and the two inference rules

$$\frac{b : \mathrm{el}(B)\ (x : \mathrm{el}(A)) \quad a\ \varepsilon\ \mathrm{el}(A)}{b\,|(x \leftarrow a) : \mathrm{el}(B)} \tag{D5.1}$$

[5]By the art of integration I mean the subject matter of, e.g., Euler, *Institutionum Calculi Integralis*. The art of substitution consists in renaming variables when necessary.

[6]Martin-Löf's calculus of explicit substitution is presented by Tasistro, 'Formulation of Martin-Löf's Type Theory with Explicit Substitutions'. Cf. § 3 of this chapter, and Abadi et al., 'Explicit substitution'.

and

$$\frac{b : \mathrm{el}(B)\ (x : \mathrm{el}(A)) \quad a = c\ \varepsilon\ \mathrm{el}(A)}{b\,|(x \leftarrow a) = b\,|(x \leftarrow c) : \mathrm{el}(B)} \tag{D5.2}$$

are meaning determining for it. Here $(x \leftarrow a)$ is called an *assignment*. Note the similarity between the definition of $b : \mathrm{el}(B)\ (x : \mathrm{el}(A))$ and the definition of $f : \mathrm{el}(A \to B)$ on p. 99; the only difference is the notation and that the variable x is mentioned in the former definition but not in the latter.

The form of assertion $b : \mathrm{el}(B)\ (x : \mathrm{el}(A))$ will now be simultaneously generalized in three directions: (1) any number of assumptions will be allowed, not just one; (2) the set over which a variable ranges may *depend* on previously introduced variables; and (3) the set B may depend on all introduced variables. For example, if we first assume that x is a number, i.e., an element of the set N, we may subsequently assume that y is a list of, e.g., Booleans, of length x, giving the assumptions

$$(x : \mathrm{el}(\mathrm{N}), y : \mathrm{el}(\mathrm{L}(\mathrm{B}, x))).$$

Such a list of assumptions will be called a *context*. From the above, it is clear that we need the two forms of assertion

$$B : \mathrm{set}\ (\Gamma)$$

and

$$\Gamma : \mathrm{context},$$

where the latter is a presupposition of the former. Examples of valid assertions of these forms should be

$$\mathrm{L}(\mathrm{B}, x) : \mathrm{set}\ (x : \mathrm{el}(\mathrm{N}))$$

and

$$(x : \mathrm{el}(\mathrm{N}), y : \mathrm{el}(\mathrm{L}(\mathrm{B}, x))) : \mathrm{context}.$$

Such a context will be called *telescopic*, cf. Figure 9.[7]

The form of assertion $B : \mathrm{set}\ (\Gamma)$ must be explained in terms of assignments of values to the sets declared in Γ. That γ is an assignments of values to the sets declared in Γ, abbreviated

$$\gamma\ \varepsilon\ \mathrm{ass}(\Gamma),$$

is a third form of assertion, with the presupposition that Γ is a context. Each of these three forms of assertion come with their corresponding assertions of equality.

Definition 12. *That Γ is a context, abbreviated Γ :* context, *means that Γ is either the empty context, written* (), *or the extension $(\Delta, x : \mathrm{el}(A))$ of a previously defined context Δ with a variable x declared to be an element of a set A in the previously defined context Δ.*

[7]This sense of the word telescopic was introduced by de Bruijn, 'Telescopic mappings in typed lambda calculus'. Note that a distinction is sometimes made between a context and a telescope, in the sense of de Bruijn: the contexts of intuitionistic type theory are *telescopic*, but they are not *telescopes*.

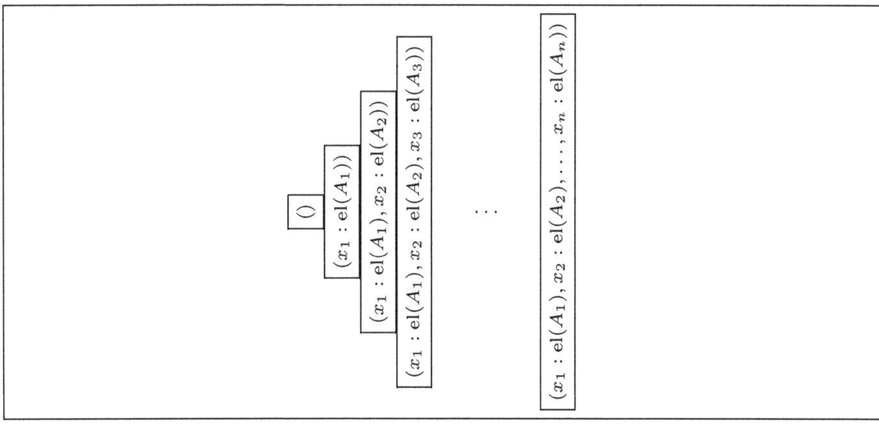

Figure 9. A context, telescopic in the sense of de Bruijn: the tubes sliding one within another are smaller and smaller contexts.

Note that there is no restriction on the variable x. That Γ is a context has no presuppositions. The axiom

$$() : \text{context} \tag{D5.3}$$

and the inference rule

$$\frac{\Delta : \text{context} \quad A : \text{set } (\Delta)}{(\Delta, x : \mathrm{el}(A)) : \text{context}} \tag{D5.4}$$

are meaning determining for the form of assertion Γ : context.

Definition 13. *That B is a set in the context Γ, abbreviated $B : \text{set } (\Gamma)$, means that, if γ is an assignment for Γ, then $B \mid \gamma$ is a noncanonical set, and that, if γ and δ are equal assignments for Γ, then $B \mid \gamma$ and $B \mid \delta$ are equal noncanonical sets.*

That B is a set in the context Γ presupposes that Γ is a context. The inference rules

$$\frac{B : \text{set } (\Gamma) \quad \gamma \; \varepsilon \; \mathrm{ass}(\Gamma)}{B \mid \gamma : \text{set}} \tag{D5.5}$$

and

$$\frac{B : \text{set } (\Gamma) \quad \gamma = \delta \; \varepsilon \; \mathrm{ass}(\Gamma)}{B \mid \gamma = B \mid \delta : \text{set}} \tag{D5.6}$$

are meaning determining for the form of assertion $B : \text{set } (\Gamma)$.

Definition 14. *That B and C are equal sets in the context Γ, abbreviated $B = C : \text{set } (\Gamma)$ means that if γ is an assignment for Γ then $B \mid \gamma$ and $C \mid \gamma$ are equal noncanonical sets.*

That B and C are equal sets in the context Γ presupposes that B and C are sets in the context Γ. The inference rule

$$\frac{B = C : \text{set } (\Gamma) \quad \gamma \; \varepsilon \; \mathrm{ass}(\Gamma)}{B \mid \gamma = C \mid \gamma : \text{set}} \tag{D5.7}$$

is meaning determining for the form of assertion $B = C : \text{set}\ (\Gamma)$. Since $B \mid \gamma$ is a new form of noncanonical set, it has to be verified that two sets of this form are equal if their parts are equal: the mediate inference rule

$$\frac{B = C : \text{set}\ (\Gamma) \quad \gamma = \delta\ \varepsilon\ \text{ass}(\Gamma)}{B \mid \gamma = C \mid \delta : \text{set}} \tag{M5.1}$$

has the schematic demonstration

$$\frac{\dfrac{B : \text{set}\ (\Gamma) \quad \gamma = \delta\ \varepsilon\ \text{ass}(\Gamma)}{B \mid \gamma = B \mid \delta : \text{set}}\ (\text{D5.6}) \qquad \dfrac{B = C : \text{set}\ (\Gamma) \quad \delta\ \varepsilon\ \text{ass}(\Gamma)}{B \mid \delta = C \mid \delta : \text{set}}\ (\text{D5.7})}{B \mid \gamma = C \mid \delta : \text{set}}\ (\text{M4.2}),$$

in which $B : \text{set}\ (\Gamma)$ is a presupposition of $B = C : \text{set}\ (\Gamma)$, i.e., an implicit premiss of (M5.1).

As expected from any notion of equality, equality between sets in a context is reflexive and cancellable, i.e., the inference rules

$$\frac{B : \text{set}\ (\Gamma)}{B = B : \text{set}\ (\Gamma)} \tag{J5.1}$$

and

$$\frac{B = D : \text{set}\ (\Gamma) \quad C = D : \text{set}\ (\Gamma)}{B = C : \text{set}\ (\Gamma)} \tag{J5.2}$$

are both valid.

Justification of (J5.1). Let the premiss be given and let γ be a given assignment for Γ. According to (D5.5), $B \mid \gamma : \text{set}$ and, by (J4.1), $B \mid \gamma = B \mid \gamma : \text{set}$, and this is what the conclusion means.

Justification of (J5.2). Let the premisses be given and let γ be a given assignment for Γ. According to (D5.7), $B \mid \gamma = D \mid \gamma : \text{set}$ and $C \mid \gamma = D \mid \gamma : \text{set}$; by (J4.2), $B \mid \gamma = C \mid \gamma : \text{set}$, and this is what the conclusion means.

As before, any notion of equality that is reflexive and cancellable is also symmetric and transitive, i.e., the mediate inference rules

$$\frac{B = C : \text{set}\ (\Gamma)}{C = B : \text{set}\ (\Gamma)} \tag{M5.2}$$

and

$$\frac{B = C : \text{set}\ (\Gamma) \quad C = D : \text{set}\ (\Gamma)}{B = D : \text{set}\ (\Gamma)} \tag{M5.3}$$

are also valid. Their schematic demonstrations are the same as those on p. 66, *mutatis mutandis.*

Definition 15. *That γ is a canonical assignment for the context Γ, abbreviated $\gamma\ \varepsilon\ \text{ass}(\Gamma)$, means that γ is the empty assignment, written $()$, if Γ is the empty context, or that γ has the form $(\delta, x \leftarrow a)$ where δ is a canonical assignment for Δ and a is a canonical element of the set which is the value of $B \mid \delta$, if Γ has the form $(\Delta, x : \text{el}(B))$.*

That γ is a canonical assignment for the context Γ presupposes that Γ is a context. The axiom

$$()\ \varepsilon\ \text{ass}() \tag{D5.8}$$

and the inference rule

$$\frac{\delta \ \varepsilon \ \mathrm{ass}(\Delta) \quad (B : \mathrm{set}\ (\Delta)) \quad B \,|\, \delta \Rightarrow A \ \varepsilon \ \mathrm{set} \quad a \ \varepsilon \ \mathrm{el}(A)}{(\delta, x \leftarrow a) \ \varepsilon \ \mathrm{ass}(\Delta, x : \mathrm{el}(B))} \tag{D5.9}$$

are meaning determining for the form of assertion $\gamma \ \varepsilon \ \mathrm{ass}(\Gamma)$.

Definition 16. *That two canonical assignments for the context Γ are equal, abbreviated $\gamma = \delta \ \varepsilon \ \mathrm{ass}(\Gamma)$, means that they have the same form and equal parts.*

That $\gamma = \delta \ \varepsilon \ \mathrm{ass}(\Gamma)$ presupposes that γ and δ are canonical assignments for the context Γ. The axiom

$$() = () \ \varepsilon \ \mathrm{ass}() \tag{D5.10}$$

and the inference rule

$$\frac{\gamma = \delta \ \varepsilon \ \mathrm{ass}(\Gamma) \quad (B : \mathrm{set}\ (\Gamma)) \quad B \,|\, \delta \Rightarrow A \ \varepsilon \ \mathrm{set} \quad a = b \ \varepsilon \ \mathrm{el}(A)}{(\gamma, x \leftarrow a) = (\delta, x \leftarrow b) \ \varepsilon \ \mathrm{ass}(\Gamma, x : \mathrm{el}(B))} \tag{D5.11}$$

are meaning determining for, and clarify the meaning of, the form of assertion $\gamma = \delta \ \varepsilon \ \mathrm{ass}(\Gamma)$. Note that the second inference rule is well-formed because $B \,|\, \gamma$ and $B \,|\, \delta$ are equal sets, whence $(\gamma, x \leftarrow a)$ is indeed a canonical assignment for $(\Gamma, x : \mathrm{el}(B))$.[8]

This notion of equality is reflexive and cancellable, i.e., the inference rules

$$\frac{\gamma \ \varepsilon \ \mathrm{ass}(\Gamma)}{\gamma = \gamma \ \varepsilon \ \mathrm{ass}(\Gamma)} \tag{J5.3}$$

and

$$\frac{\gamma = \eta \ \varepsilon \ \mathrm{ass}(\Gamma) \quad \delta = \eta \ \varepsilon \ \mathrm{ass}(\Gamma)}{\gamma = \delta \ \varepsilon \ \mathrm{ass}(\Gamma)} \tag{J5.4}$$

are both valid.

Justification of (J5.3). If Γ is the empty context, the premiss $\gamma \ \varepsilon \ \mathrm{ass}(\Gamma)$ is arrived at using (D5.8), and the conclusion follows from (D5.10). If Γ has the form $(\Delta, x : \mathrm{el}(B))$, the premiss $(\delta, x \leftarrow a) \ \varepsilon \ \mathrm{ass}(\Delta, x : \mathrm{el}(B))$ is arrived at using (D5.9) from $\delta \ \varepsilon \ \mathrm{ass}(\Delta)$, $B : \mathrm{set}\ (\Delta)$, $B \,|\, \delta \Rightarrow A \ \varepsilon$ set, and $a \ \varepsilon \ \mathrm{el}(A)$. In this case, the conclusion follows by

$$\dfrac{\dfrac{\delta \ \varepsilon \ \mathrm{ass}(\Delta)}{\delta = \delta \ \varepsilon \ \mathrm{ass}(\Delta)}\,(\mathrm{J5.3}) \quad (B : \mathrm{set}\ (\Delta)) \quad B \,|\, \delta \Rightarrow A \ \varepsilon \ \mathrm{set} \quad \dfrac{a \ \varepsilon \ \mathrm{el}(A)}{a = a \ \varepsilon \ \mathrm{el}(A)}\,(\mathrm{D3.1})}{(\delta, x \leftarrow a) = (\delta, x \leftarrow a) \ \varepsilon \ \mathrm{ass}(\Delta, x : \mathrm{el}(B))}\,(\mathrm{D5.11}).$$

Note that the appeal to (J5.3) is for the shorter context Δ.

Justification of (J5.4). If Γ is the empty context, two applications of (D5.10) followed by an application of (J5.4) can be replaced by a direct application of (D5.10). If Γ has the form $(\Delta, x : \mathrm{el}(B))$, where $B : \mathrm{set}\ (\Delta)$, the two premisses of (J5.4) have to look like

$$(\gamma, x \leftarrow a) = (\eta, x \leftarrow c) \ \varepsilon \ \mathrm{ass}(\Delta, x : \mathrm{el}(B))$$

and

$$(\delta, x \leftarrow b) = (\eta, x \leftarrow c) \ \varepsilon \ \mathrm{ass}(\Delta, x : \mathrm{el}(B)),$$

where $\gamma = \eta \ \varepsilon \ \mathrm{ass}(\Delta)$, $\delta = \eta \ \varepsilon \ \mathrm{ass}(\Delta)$, $B \,|\, \eta \Rightarrow A \ \varepsilon$ set, $a = c \ \varepsilon \ \mathrm{el}(A)$, and $b = c \ \varepsilon \ \mathrm{el}(A)$. By (J5.4), this time for the shorter context Δ, $\gamma = \delta \ \varepsilon \ \mathrm{ass}(\Delta)$; by (D3.2) $a = b \ \varepsilon \ \mathrm{el}(A)$; by (D5.6), the two sets $B \,|\, \delta$ and $B \,|\, \eta$ are equal ; if C is the value of $B \,|\, \delta$, i.e., $B \,|\, \delta \Rightarrow C \ \varepsilon$ set,

[8] Strictly speaking, the inference rules (D5.6), (D4.1), and (D3.3) are needed to show that this inference rule is well-formed.

then it follows from (D4.1) that $C = A \;\varepsilon\; \mathrm{set}$; and, by (D3.6), $a = b \;\varepsilon\; \mathrm{el}(C)$. It now follows from a final application of (D5.11) that

$$(\gamma, x \leftarrow a) = (\delta, x \leftarrow b) \;\varepsilon\; \mathrm{ass}(\Delta, x : \mathrm{el}(B)),$$

as required.

As usual, the mediate inference rules

$$\frac{\gamma = \delta \;\varepsilon\; \mathrm{ass}(\Gamma)}{\delta = \gamma \;\varepsilon\; \mathrm{ass}(\Gamma)} \tag{M5.4}$$

and

$$\frac{\gamma = \delta \;\varepsilon\; \mathrm{ass}(\Gamma) \quad \delta = \eta \;\varepsilon\; \mathrm{ass}(\Gamma)}{\gamma = \eta \;\varepsilon\; \mathrm{ass}(\Gamma)} \tag{M5.5}$$

are valid, with their usual schematic demonstrations.

Definition 17. *That b is an element of the set B in the context Γ, abbreviated $b : \mathrm{el}(B)\ (\Gamma)$, means that, if γ is an assignment for Γ, then $b \,|\, \gamma$ is a noncanonical element of the set $B \,|\, \gamma$, and that, if γ and δ are equal assignments for Γ, then $b \,|\, \gamma$ and $b \,|\, \delta$ are equal noncanonical elements of the set $B \,|\, \delta$.*

That $b : \mathrm{el}(B)\ (\Gamma)$ presupposes that B is a set in the context Γ. The two inference rules

$$\frac{b : \mathrm{el}(B)\ (\Gamma) \quad \gamma \;\varepsilon\; \mathrm{ass}(\Gamma)}{b \,|\, \gamma : \mathrm{el}(B \,|\, \gamma)} \tag{D5.12}$$

and

$$\frac{b : \mathrm{el}(B)\ (\Gamma) \quad \gamma = \delta \;\varepsilon\; \mathrm{ass}(\Gamma)}{b \,|\, \gamma = b \,|\, \delta : \mathrm{el}(B \,|\, \delta)} \tag{D5.13}$$

are meaning determining for the form of assertion $b : \mathrm{el}(B)\ (\Gamma)$.

Definition 18. *That b and c are equal elements of the set B in the context Γ, abbreviated $b = c : \mathrm{el}(B)\ (\Gamma)$, means that, if γ is an assignment for Γ, then $b \,|\, \gamma$ and $c \,|\, \gamma$ are equal elements of the set $B \,|\, \gamma$.*

That b and c are equal elements of the set B in the context Γ presupposes that b and c are elements of the set B in the context Γ. The inference rule

$$\frac{b = c : \mathrm{el}(B)\ (\Gamma) \quad \gamma \;\varepsilon\; \mathrm{ass}(\Gamma)}{b \,|\, \gamma = c \,|\, \gamma : \mathrm{el}(B \,|\, \gamma)} \tag{D5.14}$$

is meaning determining for the form of assertion $b = c \;\varepsilon\; \mathrm{el}(B)\ (\Gamma)$. Since $b \,|\, \gamma$ is a new form of noncanonical element, it has to be verified that two elements of this form are equal if their parts are equal: the mediate inference rule

$$\frac{b = c : \mathrm{el}(B)\ (\Gamma) \quad \gamma = \delta \;\varepsilon\; \mathrm{ass}(\Gamma)}{b \,|\, \gamma = c \,|\, \delta : \mathrm{el}(B \,|\, \delta)} \tag{M5.6}$$

has the schematic demonstration

$$\dfrac{\dfrac{b : \mathrm{el}(B)\ (\Gamma) \quad \gamma = \delta \;\varepsilon\; \mathrm{ass}(\Gamma)}{b \,|\, \gamma = b \,|\, \delta : \mathrm{el}(B \,|\, \delta)}\,(\mathrm{D5.13}) \quad \dfrac{b = c : \mathrm{el}(B)\ (\Gamma) \quad \delta \;\varepsilon\; \mathrm{ass}(\Gamma)}{b \,|\, \delta = c \,|\, \delta : \mathrm{el}(B \,|\, \delta)}\,(\mathrm{D5.14})}{b \,|\, \gamma = c \,|\, \delta : \mathrm{el}(B \,|\, \delta)}\,(\mathrm{M4.4}).$$

Equality between elements of a set in a context is reflexive

$$\frac{b : \mathrm{el}(B)\ (\Gamma)}{b = b : \mathrm{el}(B)\ (\Gamma)} \tag{J5.5}$$

and cancellable

$$\frac{b = d : \mathrm{el}(B)\ (\Gamma) \quad c = d : \mathrm{el}(B)\ (\Gamma)}{b = c : \mathrm{el}(B)\ (\Gamma)} \, . \tag{J5.6}$$

Justification of (J5.5). Let the premiss be given and let γ be a given canonical assignment for Γ. According to (D5.12), $b \mid \gamma : \mathrm{el}(B \mid \gamma)$; by (J4.3), $b \mid \gamma = b \mid \gamma : \mathrm{el}(B \mid \gamma)$, and this is what the conclusion means.

Justification of (J5.6). Let the premisses be given and let γ be a given canonical assignment for Γ. According to (D5.14), $b \mid \gamma = d \mid \gamma : \mathrm{el}(B \mid \gamma)$ and $c \mid \gamma = d \mid \gamma : \mathrm{el}(B \mid \gamma)$; by (J4.4), $b \mid \gamma = c \mid \gamma : \mathrm{el}(B \mid \gamma)$, and this is what the conclusion means.

As before, the mediate inference rules

$$\frac{b = c : \mathrm{el}(B)\ (\Gamma)}{c = b : \mathrm{el}(B)\ (\Gamma)} \tag{M5.7}$$

and

$$\frac{b = c : \mathrm{el}(B)\ (\Gamma) \quad c = d : \mathrm{el}(B)\ (\Gamma)}{b = d : \mathrm{el}(B)\ (\Gamma)} \tag{M5.8}$$

are valid with their usual schematic demonstrations.

Recall the rule of set conversion, (R4.1) on p. 95: a similar inference rule is valid under a context, namely, the inference rule

$$\frac{b : \mathrm{el}(B)\ (\Gamma) \quad B = C : \mathrm{set}\ (\Gamma)}{b : \mathrm{el}(C)\ (\Gamma)} \, . \tag{J5.7}$$

Justification. Let the premisses be given. There are two things to establish, corresponding to the two parts of Def. 17. First, let γ be a given canonical assignment for Γ. According to (D5.12), $b \mid \gamma : \mathrm{el}(B \mid \gamma)$, and, according to (D5.7), $B \mid \gamma = C \mid \gamma : \mathrm{set}$; by (R4.1), $b \mid \gamma : \mathrm{el}(C \mid \gamma)$. Next, let γ and δ be equal given assignments for Γ. According to (D5.13), $b \mid \gamma = b \mid \delta : \mathrm{el}(B \mid \delta)$, and, by (D5.7), $B \mid \delta = C \mid \delta : \mathrm{set}$; by (J4.5), $b \mid \gamma = b \mid \delta : \mathrm{el}(C \mid \delta)$. This completes the justification.

Similarly, set conversion holds for equal elements in a context:

$$\frac{b = c : \mathrm{el}(B)\ (\Gamma) \quad B = C : \mathrm{set}\ (\Gamma)}{b = c : \mathrm{el}(C)\ (\Gamma)} \, . \tag{J5.8}$$

Justification. Let the premisses be given and let γ be a given canonical assignment for Γ. According to (D5.14), $b \mid \gamma = c \mid \gamma : \mathrm{el}(B \mid \gamma)$, and according to (D5.7), $B \mid \gamma = C \mid \gamma : \mathrm{set}$; by (J4.5), $b \mid \gamma = c \mid \gamma : \mathrm{el}(C \mid \gamma)$.

To complete the explanation of the four forms of assertion Γ : context, B : set (Γ), $\gamma \ \varepsilon \ \mathrm{ass}(\Gamma)$, $b : \mathrm{el}(B)\ (\Gamma)$, and their corresponding equality assertions, one form of assertion remains, namely, equality between contexts.

Definition 19. *That Γ and Δ are equal contexts, abbreviated $\Gamma = \Delta \ \varepsilon$* context, *means four things: if $\gamma \ \varepsilon \ \mathrm{ass}(\Gamma)$ then $\gamma \ \varepsilon \ \mathrm{ass}(\Delta)$, if $\gamma = \delta \ \varepsilon \ \mathrm{ass}(\Gamma)$ then $\gamma = \delta \ \varepsilon \ \mathrm{ass}(\Delta)$; conversely, if $\gamma \ \varepsilon \ \mathrm{ass}(\Delta)$ then $\gamma \ \varepsilon \ \mathrm{ass}(\Gamma)$, and if $\gamma = \delta \ \varepsilon \ \mathrm{ass}(\Delta)$ then $\gamma = \delta \ \varepsilon \ \mathrm{ass}(\Gamma)$.*

This definition should be compared to Definition 5 on p. 66. The four inference rules

$$\frac{\Gamma = \Delta : \mathrm{context} \quad \gamma \ \varepsilon \ \mathrm{ass}(\Gamma)}{\gamma \ \varepsilon \ \mathrm{ass}(\Delta)} \tag{D5.15}$$

and

$$\frac{\Gamma = \Delta : \mathrm{context} \quad \gamma \ \varepsilon \ \mathrm{ass}(\Delta)}{\gamma \ \varepsilon \ \mathrm{ass}(\Gamma)} \, , \tag{D5.16}$$

and, conversely,

$$\frac{\Gamma = \Delta : \text{context} \quad \gamma = \delta \;\varepsilon\; \text{ass}(\Gamma)}{\gamma = \delta \;\varepsilon\; \text{ass}(\Delta)} \tag{D5.17}$$

and

$$\frac{\Gamma = \Delta : \text{context} \quad \gamma = \delta \;\varepsilon\; \text{ass}(\Delta)}{\gamma = \delta \;\varepsilon\; \text{ass}(\Gamma)}\,, \tag{D5.18}$$

are meaning determining for the form of assertion $\Gamma = \Delta \;\varepsilon$ context. The two inference rules

$$\frac{\Gamma : \text{context}}{\Gamma = \Gamma : \text{context}} \tag{J5.9}$$

and

$$\frac{\Gamma = \Xi : \text{context} \quad \Delta = \Xi : \text{context}}{\Gamma = \Delta : \text{context}} \tag{J5.10}$$

are justified in the same way as inference rules (J3.2) and (J3.1), *mutatis mutandis.*[9] Furthermore, the mediate inference rules

$$\frac{\Gamma = \Delta : \text{context}}{\Delta = \Gamma : \text{context}} \tag{M5.9}$$

and

$$\frac{\Gamma = \Delta : \text{context} \quad \Delta = \Xi : \text{context}}{\Gamma = \Xi : \text{context}} \tag{M5.10}$$

have their usual schematic demonstrations.

Recall that the four forms of assertion B : set (Γ), $B = C$: set (Γ), b : el(B) (Γ), and $b = c$: el(B) (Γ) are all defined in terms of canonical and equal canonical assignments for Γ. The above definition of equality between contexts is, as it were, tailor-made to make these four forms of assertion referentially transparent in Γ, i.e., to make it possible to replace Γ with any equal context and still get a valid assertion. That is, the inference rules

$$\frac{B : \text{set }(\Gamma) \quad \Gamma = \Delta : \text{context}}{B : \text{set }(\Delta)} \tag{J5.11}$$

and

$$\frac{B = C : \text{set }(\Gamma) \quad \Gamma = \Delta : \text{context}}{B = C : \text{set }(\Delta)}\,, \tag{J5.12}$$

for sets, and

$$\frac{b : \text{el}(B)\ (\Gamma) \quad \Gamma = \Delta : \text{context}}{b : \text{el}(B)\ (\Delta)} \tag{J5.13}$$

and

$$\frac{b = c : \text{el}(B)\ (\Gamma) \quad \Gamma = \Delta : \text{context}}{b = c : \text{el}(B)\ (\Delta)}\,, \tag{J5.14}$$

for elements, are all valid. These inference rules will be called rules of *context conversion.*

Justification of (J5.11). Let the premisses B : set (Γ) and $\Gamma = \Delta$: context be given. Now recall what B : set (Δ) means. Let a canonical assignment γ for the context Δ be given. By (D5.17), $\gamma \;\varepsilon\; \text{ass}(\Gamma)$, and, by (D5.5), $B \,|\, \gamma$: set. Let two equal canonical assignments γ and δ for the context Δ be given. By (D5.18), $\gamma = \delta \;\varepsilon\; \text{ass}(\Gamma)$, and, by (D5.6), $B \,|\, \gamma = B \,|\, \delta$: set. This completes the justification.

[9] These inference rules are found on p. 67.

Justification of (J5.12). Let the premisses $B = C : \text{set}\ (\Gamma)$ and $\Gamma = \Delta : \text{context}$ be given. Now recall what $B = C : \text{set}\ (\Delta)$ means. Let a canonical assignment γ for the context Δ be given. By (D5.17), $\gamma\ \varepsilon\ \text{ass}(\Gamma)$, and, by (D5.7), $B\,|\,\gamma = C\,|\,\gamma : \text{set}$. This completes the justification.

Justification of (J5.13). Let the premisses $b : \text{el}(B)\ (\Gamma)$ and $\Gamma = \Delta : \text{context}$ be given. Now recall what $b : \text{el}(B)\ (\Delta)$ means. Let a canonical assignment γ for the context Δ be given. By (D5.17), $\gamma\ \varepsilon\ \text{ass}(\Gamma)$, and, by (D5.12), $b\,|\,\gamma : \text{el}(B\,|\,\gamma)$. Let two equal canonical assignments γ and δ for the context Δ be given. By (D5.18), $\gamma = \delta\ \varepsilon\ \text{ass}(\Gamma)$, and, by (D5.13), $b\,|\,\gamma = b\,|\,\delta : \text{el}(B\,|\,\delta)$. This completes the justification.

Justification of (J5.14). Let the premisses $b = c : \text{el}(B)\ (\Gamma)$ and $\Gamma = \Delta : \text{context}$ be given. Now recall what $b = c : \text{el}(B)\ (\Delta)$ means. Let a canonical assignment γ for the context Δ be given. By (D5.17), $\gamma\ \varepsilon\ \text{ass}(\Gamma)$, and, by (D5.14), $b\,|\,\gamma = c\,|\,\gamma : \text{el}(B\,|\,\gamma)$. This completes the justification.

To complete the treatment of equal contexts, it remains to give the inference rules for *forming* equal context. Clearly, the empty context is equal to itself.

$$() = () : \text{context} \tag{J5.15}$$

Justification. The four conditions are trivially satisfied.

The inference rule

$$\frac{\Gamma = \Delta : \text{context} \quad B = C : \text{set}\ (\Gamma)}{(\Gamma, x : \text{el}(B)) = (\Delta, x : \text{el}(C)) : \text{context}} \tag{J5.16}$$

is used to form equal context extensions. Note that this inference rule is well-formed, because $B = C : \text{set}\ (\Gamma)$ presupposes that $C : \text{set}\ (\Gamma)$ which, by (J5.11), gives $C : \text{set}\ (\Delta)$; so $(\Delta, x : \text{el}(C))$ is indeed a context.

Justification. There are four things to establish, corresponding to the four parts of Definition 19. By symmetry, only the first and the third need to be given. Thus, let the premisses $\Gamma = \Delta : \text{context}$ and $B = C : \text{set}\ (\Gamma)$ be given. A canonical assignment for the context $(\Gamma, x : \text{el}(B))$ has the form $(\gamma, x \leftarrow a)$, where $\gamma\ \varepsilon\ \text{ass}(\Gamma)$ and $a\ \varepsilon\ \text{el}(A)$, for $B\,|\,\gamma \Rightarrow A\ \varepsilon\ \text{set}$. By (D5.15), $\gamma\ \varepsilon\ \text{ass}(\Delta)$; by (D5.7), $B\,|\,\gamma = C\,|\,\gamma : \text{set}$; let $C\,|\,\gamma \Rightarrow D\ \varepsilon\ \text{set}$; by (D4.1), $A = D\ \varepsilon\ \text{set}$; by (D3.3), $a\ \varepsilon\ \text{el}(D)$; and, by (D5.9), $(\gamma, x \leftarrow a)\ \varepsilon\ \text{ass}(\Delta, x : \text{el}(C))$. Two equal canonical assignments for the context $(\Gamma, x : \text{el}(B))$ have the form $(\gamma, x \leftarrow a)$ and $(\delta, x \leftarrow b)$, where $\gamma = \delta\ \varepsilon\ \text{ass}(\Gamma)$ and $a = b\ \varepsilon\ \text{el}(A)$, for $B\,|\,\delta \Rightarrow A\ \varepsilon\ \text{set}$. By (D5.16), $\gamma = \delta\ \varepsilon\ \text{ass}(\Delta)$; by (D5.7), $B\,|\,\delta = C\,|\,\delta : \text{set}$; let $C\,|\,\delta \Rightarrow D\ \varepsilon\ \text{set}$; by (D4.1), $A = D\ \varepsilon\ \text{set}$; by (D3.3), $a = b\ \varepsilon\ \text{el}(D)$; and, by (D5.11), $(\gamma, x \leftarrow a) = (\delta, x \leftarrow b)\ \varepsilon\ \text{ass}(\Delta, x : \text{el}(C))$.

This completes the explanation of the four logical categories

$$\text{context}, \qquad \text{ass}(\Gamma), \qquad \text{set}\ (\Gamma), \qquad \text{and el}(B)\ (\Gamma).$$

For the logical category $\text{ass}(\Gamma)$, a distinction is made between canonical and noncanonical assignments: canonical assignments were explained above, and the final topic of this section is noncanonical assignments.

Definition 20. *That γ is a noncanonical assignment for Γ, abbreviated $\gamma : \text{ass}(\Gamma)$, means that the value of γ is a canonical assignment for Γ. That the noncanonical assignment γ for the context Γ has the canonical assignment δ as value, abbreviated $\gamma \Rightarrow \delta\ \varepsilon\ \text{ass}(\Gamma)$, is a form of assertion defined by the computation rules which have a conclusion of this form.*

This definition should be compared to Definition 6 of noncanonical sets, on p. 91. The remarks made on the definition of noncanonical sets apply also to noncanonical assignments. That $\gamma : \text{ass}(\Gamma)$ presupposes that Γ is a context.

Definition 21. *That γ and δ are equal noncanonical assignments for the context Γ, abbreviated $\gamma = \delta : \mathrm{ass}(\Gamma)$, means that their values are equal canonical assignments for Γ.*

That $\gamma = \delta : \mathrm{ass}(\Gamma)$ presupposes that γ and δ are noncanonical assignments for Γ. The inference rule

$$\frac{\gamma = \delta : \mathrm{ass}(\Gamma) \quad \gamma \Rightarrow \varphi \;\varepsilon\; \mathrm{ass}(\Gamma) \quad \delta \Rightarrow \psi \;\varepsilon\; \mathrm{ass}(\Gamma)}{\varphi = \psi \;\varepsilon\; \mathrm{ass}(\Gamma)} \tag{D5.19}$$

is immediate from this definition and meaning determining for the form of assertion $\gamma = \delta : \mathrm{ass}(\Gamma)$.

The justifications of the inference rules

$$\frac{\gamma : \mathrm{ass}(\Gamma)}{\gamma = \gamma : \mathrm{ass}(\Gamma)} \tag{J5.17}$$

and

$$\frac{\gamma = \eta : \mathrm{ass}(\Gamma) \quad \delta = \eta : \mathrm{ass}(\Gamma)}{\gamma = \delta : \mathrm{ass}(\Gamma)} \tag{J5.18}$$

are analogous to the justifications of the inference rules (J4.1) and (J4.2).

The mediate inference rules

$$\frac{\gamma = \delta : \mathrm{ass}(\Gamma)}{\delta = \gamma : \mathrm{ass}(\Gamma)} \tag{M5.11}$$

and

$$\frac{\gamma = \delta : \mathrm{ass}(\Gamma) \quad \delta = \eta : \mathrm{ass}(\Gamma)}{\gamma = \eta : \mathrm{ass}(\Gamma)} \tag{M5.12}$$

have their usual schematic demonstrations.

§ 3. The calculus of substitutions

Recall that the γ in $\gamma \;\varepsilon\; \mathrm{el}(\Gamma)$ is called an *assignment*. The intuition behind this choice of terminology is that, when working in a context Γ, one is working, as it were, under an arbitrary assignment of values to the variables declared in Γ; this is indeed how the hypothetical forms of assertions were defined above. The components of an assignment are canonical elements of their respective sets. In this way, the hypothetical forms of assertion were explained in terms of the categorical forms of assertion.

To complete the explanation of the hypothetical forms of assertion, we also need a kind of assignment for a context Γ whose components are not objects categorically, but objects in some other context Δ. Such a generalized assignment will be called a *substitution*. I will use the form of assertion $\sigma : \Gamma \longleftarrow \Delta$, with the presuppositions that Γ and Δ are contexts, to express that σ is a substitution for the context Γ the components of which are elements in the context Δ. This form of assertion can also be read as σ being a function taking an assignment for the context Δ to an assignment for the context Γ, and this explains the arrow notation. That is, the notation $\Gamma \longleftarrow \Delta$ is a compromise between $\Gamma\ (\Delta)$ and $\Delta \longrightarrow \Gamma$. As seen from the following definition, the second explanation is the most fundamental one.

Definition 22. *That σ is a substitution from the context Δ to the context Γ, abbreviated $\sigma : \Gamma \longleftarrow \Delta$, means that if δ is a canonical assignment for Δ then $\sigma \mid \delta$ is a noncanonical assignment for Γ, and if γ and δ are equal canonical assignments for Δ then $\sigma \mid \gamma$ and $\sigma \mid \delta$ are equal noncanonical assignments for Γ.*

That σ is a substitution from Δ to Γ presupposes that Γ and Δ are context. The inference rules

$$\frac{\sigma : \Gamma \longleftarrow \Delta \quad \delta \in \mathrm{ass}(\Delta)}{\sigma \mid \delta : \mathrm{ass}(\Gamma)} \tag{D5.20}$$

and

$$\frac{\sigma : \Gamma \longleftarrow \Delta \quad \gamma = \delta \in \mathrm{ass}(\Delta)}{\sigma \mid \gamma = \sigma \mid \delta : \mathrm{ass}(\Gamma)} \tag{D5.21}$$

are meaning determining for the form of assertion $\sigma : \Gamma \longleftarrow \Delta$.

Definition 23. *That two substitutions σ and τ from the context Δ to the context Γ are equal, abbreviated $\sigma = \tau : \Gamma \longleftarrow \Delta$, means that if δ is a canonical assignment for Δ then $\sigma \mid \delta$ and $\tau \mid \delta$ are equal noncanonical assignments for Γ.*

The form of assertion $\sigma = \tau : \Gamma \longleftarrow \Delta$ presupposes that Γ and Δ are contexts. The inference rule

$$\frac{\sigma = \tau : \Gamma \longleftarrow \Delta \quad \delta \in \mathrm{ass}(\Delta)}{\sigma \mid \delta = \tau \mid \delta : \mathrm{ass}(\Gamma)} \tag{D5.22}$$

is meaning determining for the form of assertion $\sigma = \tau : \Gamma \longleftarrow \Delta$. Clearly this equality relation is reflexive

$$\frac{\sigma : \Gamma \longleftarrow \Delta}{\sigma = \sigma : \Gamma \longleftarrow \Delta}\,, \tag{J5.19}$$

and cancellable

$$\frac{\sigma = \upsilon : \Gamma \longleftarrow \Delta \quad \tau = \upsilon : \Gamma \longleftarrow \Delta}{\sigma = \tau : \Gamma \longleftarrow \Delta}\,. \tag{J5.20}$$

The justifications are analogous to those of (J4.1) and (J4.2).

Since $\sigma \mid \gamma$ is a new form of noncanonical assignment, it has to be verified that two assignments of this form are equal if their parts are equal: the mediate inference rule

$$\frac{\sigma = \tau : \Delta \longleftarrow \Gamma \quad \gamma = \delta \in \mathrm{ass}(\Gamma)}{\sigma \mid \gamma = \tau \mid \delta : \mathrm{ass}(\Delta)} \tag{M5.13}$$

has the schematic demonstration

$$\frac{\dfrac{\sigma : \Delta \longleftarrow \Gamma \quad \gamma = \delta \in \mathrm{ass}(\Gamma)}{\sigma \mid \gamma = \sigma \mid \delta : \mathrm{ass}(\Delta)}\,(\text{D5.21}) \qquad \dfrac{\sigma = \tau : \Delta \longleftarrow \Gamma \quad \delta \in \mathrm{ass}(\Gamma)}{\sigma \mid \delta = \tau \mid \delta : \mathrm{ass}(\Delta)}\,(\text{D5.22})}{\sigma \mid \gamma = \tau \mid \delta : \mathrm{ass}(\Delta)}\,(\text{M5.12}).$$

The two inference rules

$$\frac{\sigma : \Gamma \longleftarrow \Delta \quad \Delta = \Xi : \mathrm{context}}{\sigma : \Gamma \longleftarrow \Xi} \tag{J5.21}$$

and

$$\frac{\sigma = \tau : \Gamma \longleftarrow \Delta \quad \Delta = \Xi : \mathrm{context}}{\sigma = \tau : \Gamma \longleftarrow \Xi} \tag{J5.22}$$

have justifications similar to those of (J5.11) and (J5.12).

Composition of two substitutions σ and τ is denoted by a circle, like function composition. The inference rule

$$\frac{\sigma : \Gamma \longleftarrow \Delta \quad \tau : \Delta \longleftarrow \Phi}{\sigma \circ \tau : \Gamma \longleftarrow \Phi} \tag{R5.1}$$

is recognized as valid in virtue of the computation rule

$$\frac{\tau \,|\, \varphi \Rightarrow \delta \;\varepsilon\; \mathrm{ass}(\Delta) \quad \sigma \,|\, \delta \Rightarrow \gamma \;\varepsilon\; \mathrm{ass}(\Gamma)}{\sigma \circ \tau \,|\, \varphi \Rightarrow \gamma \;\varepsilon\; \mathrm{ass}(\Gamma)} \,. \tag{C5.1}$$

In this computation rule, the premisses of (R5.1) are not written out again, since the two inference rules, (C5.1) and (R5.1), are to be understood together; moreover, the premiss that $\varphi \;\varepsilon\; \mathrm{ass}(\Phi)$ is also suppressed, since for this computation rule to make sense, φ has to be a canonical assignment for the context Φ.

As for any new form of expression, it has to be verified that two terms of $\circ$ form are equal if their parts are equal, i.e., that the inference rule

$$\frac{\sigma = \tau : \Gamma \longleftarrow \Delta \quad \zeta = \eta : \Delta \longleftarrow \Phi}{\sigma \circ \zeta = \tau \circ \eta : \Gamma \longleftarrow \Phi} \tag{J5.23}$$

is valid.

Justification. Let the premisses $\sigma = \tau : \Gamma \longleftarrow \Delta$ and $\zeta = \eta : \Delta \longleftarrow \Phi$ be given. The conclusion means that $\sigma \circ \zeta \,|\, \varphi = \tau \circ \eta \,|\, \varphi : \mathrm{ass}(\Gamma)$, where φ is a given canonical assignment for Φ. Let the computation trace of $\sigma \circ \zeta \,|\, \varphi$ be

$$\frac{\zeta \,|\, \varphi \Rightarrow \delta_1 \;\varepsilon\; \mathrm{ass}(\Delta) \quad \sigma \,|\, \delta_1 \Rightarrow \gamma_1 \;\varepsilon\; \mathrm{ass}(\Gamma)}{\sigma \circ \zeta \,|\, \varphi \Rightarrow \gamma_1 \;\varepsilon\; \mathrm{ass}(\Gamma)} \,,$$

and that of $\tau \circ \eta \,|\, \varphi$ be

$$\frac{\eta \,|\, \varphi \Rightarrow \delta_2 \;\varepsilon\; \mathrm{ass}(\Delta) \quad \tau \,|\, \delta_2 \Rightarrow \gamma_2 \;\varepsilon\; \mathrm{ass}(\Gamma)}{\tau \circ \eta \,|\, \varphi \Rightarrow \gamma_2 \;\varepsilon\; \mathrm{ass}(\Gamma)} \,.$$

It remains to show that $\gamma_1 = \gamma_2 \;\varepsilon\; \mathrm{ass}(\Gamma)$:

$$\frac{\sigma = \tau : \Gamma \longleftarrow \Delta \qquad \dfrac{\dfrac{\zeta = \eta : \Delta \longleftarrow \Phi \quad \varphi \;\varepsilon\; \mathrm{ass}(\Phi)}{\zeta \,|\, \varphi = \eta \,|\, \varphi : \Delta} \quad \begin{array}{c}\zeta \,|\, \varphi \Rightarrow \delta_1 \;\varepsilon\; \mathrm{ass}(\Delta) \\ \eta \,|\, \varphi \Rightarrow \delta_2 \;\varepsilon\; \mathrm{ass}(\Delta)\end{array}}{\delta_1 = \delta_2 \;\varepsilon\; \mathrm{ass}(\Delta)}}{\sigma \,|\, \delta_1 = \tau \,|\, \delta_2 : \mathrm{ass}(\Gamma)} \quad \begin{array}{c}\sigma \,|\, \delta_1 \Rightarrow \gamma_1 \;\varepsilon\; \mathrm{ass}(\Gamma) \\ \tau \,|\, \delta_2 \Rightarrow \gamma_2 \;\varepsilon\; \mathrm{ass}(\Gamma)\end{array}$$
$$\overline{\qquad \gamma_1 = \gamma_2 \;\varepsilon\; \mathrm{ass}(\Gamma) \qquad} \,.$$

This completes the justification.

Composition is associative, i.e., the inference rule

$$\frac{\sigma : \Gamma \longleftarrow \Delta \quad \tau : \Delta \longleftarrow \Phi \quad \upsilon : \Phi \longleftarrow \Lambda}{(\sigma \circ \tau) \circ \upsilon = \sigma \circ (\tau \circ \upsilon) : \Gamma \longleftarrow \Lambda} \tag{J5.24}$$

is valid.

Justification. Let the premisses $\sigma : \Gamma \longleftarrow \Delta$, $\tau : \Delta \longleftarrow \Phi$, and $\upsilon : \Phi \longleftarrow \Lambda$ be given, and let λ be a given canonical assignment for the context Λ. Compare the two computation traces

$$\frac{\upsilon \,|\, \lambda \Rightarrow \varphi \;\varepsilon\; \mathrm{ass}(\Phi) \qquad \dfrac{\tau \,|\, \varphi \Rightarrow \delta \;\varepsilon\; \mathrm{ass}(\Delta) \quad \sigma \,|\, \delta \Rightarrow \gamma \;\varepsilon\; \mathrm{ass}(\Gamma)}{\sigma \circ \tau \,|\, \varphi \Rightarrow \gamma \;\varepsilon\; \mathrm{ass}(\Gamma)}}{(\sigma \circ \tau) \circ \upsilon \,|\, \lambda \Rightarrow \gamma \;\varepsilon\; \mathrm{ass}(\Gamma)}$$

and

$$\cfrac{\cfrac{\upsilon \mid \lambda \Rightarrow \varphi \ \varepsilon \ \mathrm{ass}(\Phi) \quad \tau \mid \varphi \Rightarrow \delta \ \varepsilon \ \mathrm{ass}(\Delta)}{\tau \circ \upsilon \mid \lambda \Rightarrow \delta \ \varepsilon \ \mathrm{ass}(\Delta)} \qquad \sigma \mid \delta \Rightarrow \gamma \ \varepsilon \ \mathrm{ass}(\Gamma)}{\sigma \circ (\tau \circ \upsilon) \mid \lambda \Rightarrow \gamma \ \varepsilon \ \mathrm{ass}(\Gamma)} \ .$$

They show that the values of $(\sigma \circ \tau) \circ \upsilon \mid \lambda$ and $\sigma \circ (\tau \circ \upsilon) \mid \lambda$ are equal canonical assignments, in fact the same canonical assignment γ, for the context Γ for the arbitrarily given canonical assignment λ for the context Λ. This is precisely what the equality in the conclusion of (J5.24) means.

As a first example of a substitution, we have the identity substitution, abbreviated id, from any context to itself. The inference rule

$$\frac{(\Gamma : \mathrm{context})}{\mathrm{id} : \Gamma \longleftarrow \Gamma} \tag{R5.2}$$

is recognized as valid in virtue of the computation rule

$$\mathrm{id} \mid \gamma \Rightarrow \gamma \ \varepsilon \ \mathrm{ass}(\Gamma). \tag{C5.2}$$

As usual, the premiss of (R5.2) is put within parentheses to show that it is needed only as a presupposition of the conclusion.

The identity substitution is a left identity with respect to composition of substitution:

$$\frac{\sigma : \Gamma \longleftarrow \Delta}{\mathrm{id} \circ \sigma = \sigma : \Gamma \longleftarrow \Delta} \ . \tag{J5.25}$$

Justification. Let the premiss $\sigma : \Gamma \longleftarrow \Delta$ be given and let δ be an arbitrarily given canonical assignment for Δ. Compare the computation trace

$$\frac{\sigma \mid \delta \Rightarrow \gamma \ \varepsilon \ \mathrm{ass}(\Gamma) \quad \mathrm{id} \mid \gamma \Rightarrow \gamma \ \varepsilon \ \mathrm{ass}(\Gamma)}{\mathrm{id} \circ \sigma \mid \delta \Rightarrow \gamma \ \varepsilon \ \mathrm{ass}(\Gamma)}$$

for $\mathrm{id} \circ \sigma \mid \delta$ to the computation trace

$$\sigma \mid \delta \Rightarrow \gamma \ \varepsilon \ \mathrm{ass}(\Gamma),$$

for $\sigma \mid \delta$. Since $\mathrm{id} \mid \gamma \Rightarrow \gamma \ \varepsilon \ \mathrm{ass}(\Gamma)$ is a computation rule without premisses, it follows that $\mathrm{id} \circ \sigma$ is equal to σ as a substitution $\Gamma \longleftarrow \Delta$.

The identity substitution is a right identity with respect to composition of substitution:

$$\frac{\sigma : \Gamma \longleftarrow \Delta}{\sigma \circ \mathrm{id} = \sigma : \Gamma \longleftarrow \Delta} \ . \tag{J5.26}$$

The justification is analogous to the previous justification.

A substitution $\sigma : \Gamma \longleftarrow \Delta$ can also be viewed as a *translation* that translates an assertion made in the context Γ into a corresponding assertion made in the context Δ. This understanding underlies the inference rules

$$\frac{A : \mathrm{set}\ (\Gamma) \quad \sigma : \Gamma \longleftarrow \Delta}{A \circ \sigma : \mathrm{set}\ (\Delta)} \tag{R5.3}$$

and

$$\frac{a : \mathrm{el}(A)\ (\Gamma) \quad \sigma : \Gamma \longleftarrow \Delta}{a \circ \sigma : \mathrm{el}(A \circ \sigma)\ (\Delta)} \ . \tag{R5.4}$$

The corresponding computation rules are given by

$$\frac{\sigma \mid \delta \Rightarrow \gamma \ \varepsilon \ \mathrm{ass}(\Gamma) \quad A \mid \gamma \Rightarrow B \ \varepsilon \ \mathrm{set}}{A \circ \sigma \mid \delta \Rightarrow B \ \varepsilon \ \mathrm{set}} \tag{C5.3}$$

and

$$\frac{\sigma \,|\, \delta \Rightarrow \gamma \;\varepsilon\; \mathrm{ass}(\Gamma) \quad \mathrm{el}(A \,|\, \gamma) : a \,|\, \gamma \Rightarrow b \;\varepsilon\; \mathrm{el}(B)}{\mathrm{el}(A \circ \sigma \,|\, \delta) : a \circ \sigma \,|\, \delta \Rightarrow b \;\varepsilon\; \mathrm{el}(B)} \,. \tag{C5.4}$$

Note that σ, viewed as a *function* from Δ to Γ, becomes a *translation* of assertions made in the context Γ to assertions made in the context Δ.

Now $A \circ \sigma$ and $a \circ \sigma$ are two new forms of expression, and it has to be verified that two expressions of one of these forms are equal if their parts are equal, i.e., that the inference rules

$$\frac{A = B : \mathrm{set}\ (\Gamma) \quad \sigma = \tau : \Gamma \longleftarrow \Delta}{A \circ \sigma = B \circ \tau : \mathrm{set}\ (\Delta)} \tag{J5.27}$$

and

$$\frac{a = b : \mathrm{el}(A)\ (\Gamma) \quad \sigma = \tau : \Gamma \longleftarrow \Delta}{a \circ \sigma = b \circ \tau : \mathrm{el}(A \circ \tau)\ (\Delta)} \tag{J5.28}$$

are valid. The justifications are analogous to that of (J5.23).

Composing a set, or an element of a set, with the identity substitution yields the same set or element, i.e., the inference rules

$$\frac{B : \mathrm{set}\ (\Gamma)}{B \circ \mathrm{id} = B : \mathrm{set}\ (\Gamma)} \tag{J5.29}$$

and

$$\frac{b : \mathrm{el}(B)\ (\Gamma)}{b \circ \mathrm{id} = b : \mathrm{el}(B)\ (\Gamma)} \tag{J5.30}$$

are both valid. Note that the latter inference rule is well-formed because of the former and (J5.7). Again, these inference rules are justified by comparing the computation traces of the two sides of the equality under an arbitrarily given assignment for the context Γ.

Inference rule (J5.24) is valid also for sets and elements, as expressed by the inference rules

$$\frac{A : \mathrm{set}\ (\Gamma) \quad \sigma : \Gamma \longleftarrow \Delta \quad \tau : \Delta \longleftarrow \Phi}{(A \circ \sigma) \circ \tau = A \circ (\sigma \circ \tau) : \mathrm{set}\ (\Phi)} \tag{J5.31}$$

and

$$\frac{a : \mathrm{el}(A)\ (\Gamma) \quad \sigma : \Gamma \longleftarrow \Delta \quad \tau : \Delta \longleftarrow \Phi}{(a \circ \sigma) \circ \tau = a \circ (\sigma \circ \tau) : \mathrm{el}(A \circ \sigma \circ \tau)\ (\Phi)} \,. \tag{J5.32}$$

The justifications are analogous to that of (J5.24).

The two most important forms of substitution are the empty substitution, written $()$, and the extension substitution $(\sigma, x \leftarrow b)$ of a substitution σ with an element b. These substitutions are direct generalizations of the corresponding assignments. The empty substitution is a substitution from any context Γ into the empty context:

$$\frac{(\Gamma : \mathrm{context})}{() : () \longleftarrow \Gamma} \tag{R5.5}$$

This inference rule is recognized as valid in virtue of the computation rule

$$() \,|\, \gamma \Rightarrow () \;\varepsilon\; \mathrm{ass}(). \tag{C5.5}$$

Composing the empty substitution with any substitution gives the empty substitution as result, i.e., the inference rule

$$\frac{\sigma : \Gamma \longleftarrow \Delta}{() \circ \sigma = () : () \longleftarrow \Delta} \tag{J5.33}$$

is valid.

Justification. Let the premiss $\sigma : \Gamma \longleftarrow \Delta$ be given and let δ be a given substitution for Δ. Compare the computation trace

$$\frac{\sigma \,|\, \delta \Rightarrow \gamma \;\varepsilon\; \mathrm{ass}(\Gamma) \quad () \,|\, \gamma \Rightarrow () \;\varepsilon\; \mathrm{ass}()}{() \circ \sigma \,|\, \delta \Rightarrow () \;\varepsilon\; \mathrm{ass}()}$$

for $() \circ \sigma$ to the computation trace $() \,|\, \delta \Rightarrow () \;\varepsilon\; \mathrm{ass}()$. Since the result is $()$ in both cases, $() \circ \sigma$ and $()$ are equal substitutions from Δ to the empty context.

The extension of a substitution is formed according to the inference rule

$$\frac{\sigma : \Gamma \longleftarrow \Delta \quad (B : \mathrm{set}\ (\Gamma)) \quad b : \mathrm{el}(B \circ \sigma)\ (\Delta)}{(\sigma, x \leftarrow b) : (\Gamma, x : \mathrm{el}(B)) \longleftarrow \Delta}\ . \tag{R5.6}$$

Note that $b : \mathrm{el}(B \circ \sigma)\ (\Delta)$. The extension substitution is computed according to the computation rule

$$\frac{\sigma \,|\, \delta \Rightarrow \gamma \;\varepsilon\; \mathrm{ass}(\Gamma) \quad \mathrm{el}(B \circ \sigma \,|\, \delta) : b \,|\, \delta \Rightarrow c \;\varepsilon\; \mathrm{el}(C)}{(\sigma, x \leftarrow b) \,|\, \delta \Rightarrow (\gamma, x \leftarrow c) \;\varepsilon\; \mathrm{ass}(\Gamma, x : \mathrm{el}(B))}\ . \tag{C5.6}$$

Observe that $B \circ \sigma \,|\, \delta$ and $B \,|\, \gamma$ are equal sets, so $(\gamma, x \leftarrow c)$ is indeed an assignment for the context $(\Gamma, x : \mathrm{el}(B))$. As always, it has to be verified that two terms of the form $(\sigma, x \leftarrow b)$ are equal if their parts are equal.

$$\frac{\sigma = \tau : \Gamma \longleftarrow \Delta \quad (B : \mathrm{set}\ (\Gamma)) \quad b = c : \mathrm{el}(B \circ \tau)\ (\Delta)}{(\sigma, x \leftarrow b) = (\tau, x \leftarrow c) : (\Gamma, x : \mathrm{el}(B)) \longleftarrow \Delta} \tag{J5.34}$$

Justification. Let the premisses $\sigma = \tau : \Gamma \longleftarrow \Delta$, $B : \mathrm{set}\ (\Gamma)$, and $b = c : \mathrm{el}(B \circ \tau)\ (\Delta)$ be given, and let δ be an arbitrarily given canonical assignment for Δ. Furthermore, let γ_1, γ_2, d_1, and d_2 be defined according to the computation traces

$$\frac{\sigma \,|\, \delta \Rightarrow \gamma_1 \;\varepsilon\; \mathrm{ass}(\Gamma) \quad \mathrm{el}(B \circ \sigma \,|\, \delta) : b \,|\, \delta \Rightarrow d_1 \;\varepsilon\; \mathrm{el}(C_1)}{(\sigma, x \leftarrow b) \,|\, \delta \Rightarrow (\gamma_1, x \leftarrow d_1) \;\varepsilon\; (\Gamma, x : \mathrm{el}(B))}$$

and

$$\frac{\tau \,|\, \delta \Rightarrow \gamma_2 \;\varepsilon\; \mathrm{ass}(\Gamma) \quad \mathrm{el}(B \circ \tau \,|\, \delta) : c \,|\, \delta \Rightarrow d_2 \;\varepsilon\; \mathrm{el}(C_2)}{(\tau, x \leftarrow c) \,|\, \delta \Rightarrow (\gamma_2, x \leftarrow d_2) \;\varepsilon\; (\Gamma, x : \mathrm{el}(B))}\ .$$

It remains to show that $(\gamma_1, x \leftarrow d_1)$ and $(\gamma_2, x \leftarrow d_2)$ are equal canonical assignments for $(\Gamma, x : \mathrm{el}(B))$. First:

$$\cfrac{\cfrac{\sigma = \tau : \Gamma \longleftarrow \Delta \quad \delta \;\varepsilon\; \mathrm{ass}(\Delta)}{\sigma \,|\, \delta = \tau \,|\, \delta : \Gamma}\ \text{(D5.22)} \qquad \sigma \,|\, \delta \Rightarrow \gamma_1 \;\varepsilon\; \mathrm{ass}(\Gamma) \quad \tau \,|\, \delta \Rightarrow \gamma_2 \;\varepsilon\; \mathrm{ass}(\Gamma)}{\gamma_1 = \gamma_2 \;\varepsilon\; \mathrm{ass}(\Gamma)}\ \text{(D5.19)}.$$

Next:

$$\cfrac{\cfrac{b = c : \mathrm{el}(B \circ \tau)\ (\Delta) \quad \delta \;\varepsilon\; \mathrm{ass}(\Delta)}{b \,|\, \delta = c \,|\, \delta : \mathrm{el}(B \circ \tau \,|\, \delta)}\ \text{(D5.14)} \qquad \begin{array}{l} \mathrm{el}(B \circ \tau \,|\, \delta) : b \,|\, \delta \Rightarrow d_1 \;\varepsilon\; \mathrm{el}(C_2) \\ \mathrm{el}(B \circ \tau \,|\, \delta) : c \,|\, \delta \Rightarrow d_2 \;\varepsilon\; \mathrm{el}(C_2) \end{array}}{d_1 = d_2 \;\varepsilon\; \mathrm{el}(C_2)}\ \text{(D4.2)}.$$

The conclusion follows from (D5.11).

The next inference rule shows that a substitution τ can be moved inside a substitution of the form $(\sigma, x \leftarrow b)$.

$$\frac{\sigma : \Gamma \longleftarrow \Delta \quad (B : \text{set}\ (\Gamma)) \quad b : \text{el}(B \circ \sigma)\ (\Delta) \quad \tau : \Delta \longleftarrow \Phi}{(\sigma, x \leftarrow b) \circ \tau = (\sigma \circ \tau, x \leftarrow b \circ \tau) : (\Gamma, x : \text{el}(B)) \longleftarrow \Phi} \,. \tag{J5.35}$$

Justification. Let φ be an arbitrary assignment for Φ and compare the two computation traces

$$\frac{\tau \mid \varphi \Rightarrow \delta \;\varepsilon\; \text{ass}(\Delta) \qquad \dfrac{\sigma \mid \delta \Rightarrow \gamma \;\varepsilon\; \text{ass}(\Gamma) \quad \text{el}(B \circ \sigma \mid \delta) : b \mid \delta \Rightarrow c \;\varepsilon\; \text{el}(C)}{(\sigma, x \leftarrow b) \mid \delta \Rightarrow (\gamma, x \leftarrow c) \;\varepsilon\; \text{ass}(\Gamma, x : \text{el}(B))}}{(\sigma, x \leftarrow b) \circ \tau \mid \varphi \Rightarrow (\gamma, x \leftarrow c) \;\varepsilon\; \text{ass}(\Gamma, x : \text{el}(B))}$$

and

$$\frac{\dfrac{\tau \mid \varphi \Rightarrow \delta \;\varepsilon\; \text{ass}(\Delta) \quad \sigma \mid \delta \Rightarrow \gamma \;\varepsilon\; \text{ass}(\Gamma)}{(\sigma \circ \tau) \mid \varphi \Rightarrow \gamma \;\varepsilon\; \text{ass}(\Gamma)} \qquad \dfrac{\begin{array}{c}\tau \mid \varphi \Rightarrow \delta \;\varepsilon\; \text{ass}(\Delta) \\ \text{el}(B \circ \sigma \mid \delta) : b \mid \delta \Rightarrow c \;\varepsilon\; \text{el}(C)\end{array}}{\text{el}(B \circ \sigma \circ \tau \mid \varphi) : b \circ \tau \mid \varphi \Rightarrow c \;\varepsilon\; \text{el}(C)}}{(\sigma \circ \tau, x \leftarrow b \circ \tau) \mid \varphi \Rightarrow (\gamma, x \leftarrow c) \;\varepsilon\; \text{ass}(\Gamma, x : \text{el}(B)).}$$

They show that $(\sigma, x \leftarrow b) \circ \tau \mid \varphi$ and $(\sigma \circ \tau, x \leftarrow b \circ \tau) \mid \varphi$ indeed are equal assignments for the context $(\Gamma, x : \text{el}(B))$; their common value depends only on the values of $\tau \mid \varphi : \text{ass}(\Delta)$, $\sigma \mid \delta : \text{ass}(\Gamma)$, and $b \mid \delta : \text{el}(B \circ \sigma \mid \delta)$.

It remains to treat of *weakening* and *assumption.* As of yet, variables have only occurred in contexts, assignments, and substitutions; but the whole purpose of using variables is that, if they are *assumed* to be elements of a certain set, then they *are* elements of that set. The following inference rule is called the *assumption rule*:

$$\frac{B : \text{set}\ (\Gamma)}{x : \text{el}(B)\ (\Gamma, x : \text{el}(B))} \,.$$

Unfortunately, with the present understanding of hypothetical assertions, this inference rule is not even well-formed, because the conclusion presupposes that B is a set in the context $(\Gamma, x : \text{el}(B))$ while, according to the premiss, B is only a set in the context Γ. This is why *weakening* has to be treated of before assumption. The present formulation of intuitionistic type theory uses *explicit weakening*, i.e., there is a special weakening substitution:[10]

$$\frac{B : \text{set}\ (\Gamma)}{\text{p} : \Gamma \longleftarrow (\Gamma, x : \text{el}(B))} \tag{R5.7}$$

This inference rule is recognized as valid in virtue of the computation rule

$$\text{p} \mid (\gamma, x \leftarrow a) \Rightarrow \gamma \;\varepsilon\; \text{ass}(\Gamma), \tag{C5.7}$$

in which γ is an arbitrarily given canonical assignment for Γ and a is an arbitrarily given canonical element of the canonical set that is the value of $B \mid \gamma$, so that $(\gamma, x \leftarrow a)$ is an arbitrarily given canonical assignment for the context $(\Gamma, x : \text{el}(B))$. The following justified inference rule shows that the

[10]The notation p is taken from Hofmann, 'Syntax and Semantics of Dependent Types', § 2.4–§ 3.3. As pointed out to me by Prof. Palmgren, my approach to explicit substitution is similar to Hofmann's category theoretic approach, and I have adopted some of his notation. Instead of Hofmann's notation q for "the variable", I use ordinary variables, but, if all variables are named q, my notation agrees with his. Further references on the category theoretic approach to the semantics of type theory are: Cartmell, 'Generalized algebraic theories and contextual categories'; Dybjer, 'Internal type theory'; and Jacobs, *Categorical Logic and Type Theory*, Ch. 10.

weakening substitution p composed with a substitution of the form $(\sigma, x \leftarrow b)$ is equal to σ:

$$\frac{\sigma : \Gamma \longleftarrow \Delta \quad (B : \mathrm{set}\,(\Gamma)) \quad b : \mathrm{el}(B \circ \sigma)\,(\Delta)}{\mathrm{p} \circ (\sigma, x \leftarrow b) = \sigma : \Gamma \longleftarrow \Delta} \tag{J5.36}$$

Justification. Let the premisses $\sigma : \Gamma \longleftarrow \Delta$, $B : \mathrm{set}\,(\Gamma)$, and $b : \mathrm{el}(B \circ \sigma)\,(\Delta)$ be given; and let δ be an arbitrarily given assignment for Δ. If the value of $\sigma \mid \delta$ is the canonical assignment γ for the context Γ, then the value of $\mathrm{p} \circ (\sigma, x \leftarrow b) \mid \delta$ is also γ, as shown by the computation trace

$$\frac{\dfrac{\sigma \mid \delta \Rightarrow \gamma \;\varepsilon\; \mathrm{ass}(\Gamma) \quad \mathrm{el}(B \circ \sigma \mid \delta) : b \mid \delta \Rightarrow a \;\varepsilon\; \mathrm{el}(A)}{(\sigma, x \leftarrow b) \mid \delta \Rightarrow (\gamma, x \leftarrow a) \;\varepsilon\; \mathrm{ass}(\Gamma, x : \mathrm{el}(B))} \qquad \mathrm{p} \mid (\gamma, x \leftarrow a) \Rightarrow \gamma \;\varepsilon\; \mathrm{ass}(\Gamma)}{\mathrm{p} \circ (\sigma, x \leftarrow b) \mid \delta \Rightarrow \gamma \;\varepsilon\; \mathrm{ass}(\Gamma)}\,.$$

This completes the justification.

The *assumption rule* with explicit weakening now becomes

$$\frac{B : \mathrm{set}\,(\Gamma)}{x : \mathrm{el}(B \circ \mathrm{p})\,(\Gamma, x : \mathrm{el}(B))}\,. \tag{R5.8}$$

This inference rule is recognized as valid in virtue of the computation rule

$$\mathrm{el}(B \circ \mathrm{p} \mid (\gamma, x \leftarrow a)) : x \mid (\gamma, x \leftarrow a) \Rightarrow a \;\varepsilon\; \mathrm{el}(A) \tag{C5.8}$$

in which γ is an arbitrarily given canonical assignment for Γ and a is arbitrarily given canonical element of the canonical set that is the value of $B \mid \gamma$. Note that $B \circ \mathrm{p} \mid (\gamma, x \leftarrow a)$ and $B \mid \gamma$ are equal sets and that a is, by assumption, a canonical element of the set A which is the value of $B \mid \gamma$. Furthermore, there is no freshness condition on x, i.e., several variables in the context may have the same name x, and x refers to the rightmost of them, $x \circ \mathrm{p}$ to the next, etc. The composition $x \circ (\sigma, x \leftarrow b)$ is equal to b:

$$\frac{\sigma : \Gamma \longleftarrow \Delta \quad (B : \mathrm{set}\,(\Gamma)) \quad b : \mathrm{el}(B \circ \sigma)\,(\Delta)}{x \circ (\sigma, x \leftarrow b) = b : \mathrm{el}(B \circ \sigma)\,(\Delta)} \tag{J5.37}$$

This inference rule is well-formed because, first,

$$\frac{\dfrac{B : \mathrm{set}\,(\Gamma)}{x : \mathrm{el}(B \circ \mathrm{p})\,(\Gamma, x : \mathrm{el}(B))} \quad \dfrac{\sigma : \Gamma \longleftarrow \Delta \quad b : \mathrm{el}(B \circ \sigma)\,(\Delta)}{(\sigma, x \leftarrow b) : (\Gamma, x : \mathrm{el}(B)) \longleftarrow \Delta}}{x \circ (\sigma, x \leftarrow b) : \mathrm{el}((B \circ \mathrm{p}) \circ (\sigma, x \leftarrow b))\,(\Delta)}\,,$$

and, next,

$$(B \circ \mathrm{p}) \circ (\sigma, x \leftarrow b) = B \circ (\mathrm{p} \circ (\sigma, x \leftarrow b)) = B \circ \sigma : \mathrm{set}\,(\Delta),$$

using (J5.31), (J5.36), and (M5.3); finally, (J5.7) gives that $x \circ (\sigma, x \leftarrow b) : \mathrm{el}(B \circ \sigma)\,(\Delta)$, as required.

Justification of (J5.37). Let the premisses be given and let δ be an arbitrarily given assignment for Δ. Let $b \mid \delta$ have the value a, i.e.,

$$\mathrm{el}(B \circ \sigma \mid \delta) : b \mid \delta \Rightarrow a \;\varepsilon\; \mathrm{el}(A).$$

The value of $x \circ (\sigma, x \leftarrow b) \mid \delta$ is also a, as shown by

$$\frac{\dfrac{\sigma \mid \delta \Rightarrow \gamma \;\varepsilon\; \mathrm{ass}(\Gamma) \quad \mathrm{el}(B \circ \sigma \mid \delta) : b \mid \delta \Rightarrow a \;\varepsilon\; \mathrm{el}(A)}{(\sigma, x \leftarrow b) \mid \delta \Rightarrow (\gamma, x \leftarrow a) \;\varepsilon\; \mathrm{ass}(\Gamma, x : \mathrm{el}(B))} \quad \begin{array}{l} B \circ \mathrm{p} \mid (\gamma, x \leftarrow a) : \\ x \mid (\gamma, x \leftarrow a) \Rightarrow a \;\varepsilon\; \mathrm{el}(A) \end{array}}{\mathrm{el}(B \circ \sigma \mid \delta) : x \circ (\sigma, x \leftarrow b) \mid \delta \Rightarrow a \;\varepsilon\; \mathrm{el}(A)}\,.$$

This completes the justification.

At this point, it can be noted that the identity substitution, id, always can be eliminated, i.e., it is always expressible in terms of other substitutions according to the rules

$$\mathrm{id} = () : () \longleftarrow () \tag{J5.38}$$

and

$$\frac{B : \mathrm{set}\ (\Gamma)}{\mathrm{id} = (\mathrm{p}, x \leftarrow x) : (\Gamma, x : \mathrm{el}(B)) \longleftarrow (\Gamma, x : \mathrm{el}(B))}\ . \tag{J5.39}$$

Justification of (J5.38). Since the empty substitution is the only substitution for the empty context, it suffices to compare the computation

$$\mathrm{id}\,|\,() \Rightarrow ()\ \varepsilon\ \mathrm{ass}()$$

to the computation

$$()\,|\,() \Rightarrow ()\ \varepsilon\ \mathrm{ass}().$$

The result, i.e., the empty substitution, (), is the same in both cases.

Justification of (J5.39). Let the premiss $B : \mathrm{set}\ (\Gamma)$ be given and let $(\gamma, x \leftarrow a)$ be an arbitrarily given assignment for $(\Gamma, x : \mathrm{el}(B))$, i.e., with $\gamma\ \varepsilon\ \mathrm{ass}(\Gamma)$ and $a\ \varepsilon\ \mathrm{el}(A)$, where $B\,|\,\gamma \Rightarrow A\ \varepsilon\ \mathrm{set}$. The left-hand side of the equality is computed by

$$\mathrm{id}\,|(\gamma, x \leftarrow a) \Rightarrow (\gamma, x \leftarrow a)\ \varepsilon\ \mathrm{ass}(\Gamma, x : \mathrm{el}(B))$$

when applied to this assignment, and the right-hand side by

$$\frac{\mathrm{p}\,|(\gamma, x \leftarrow a) \Rightarrow \gamma\ \varepsilon\ \mathrm{ass}(\Gamma) \quad \mathrm{el}(B\,|\,\gamma) : x\,|(\gamma, x \leftarrow a) \Rightarrow a\ \varepsilon\ \mathrm{el}(A)}{(\mathrm{p}, x \leftarrow x)\,|(\gamma, x \leftarrow a) \Rightarrow (\gamma, x \leftarrow a)\ \varepsilon\ \mathrm{ass}(\Gamma, x : \mathrm{el}(B))}\ .$$

The result is the same in both cases.

If the context Γ is empty, the substitution $\mathrm{p} : \Gamma \longleftarrow (\Gamma, x : \mathrm{el}(B))$ is equal to the empty substitution:

$$\mathrm{p} = \mathrm{id} \circ \mathrm{p} = () \circ \mathrm{p} = () : () \longleftarrow (x : \mathrm{el}(B)),$$

and, if Γ is the context $(\Delta, y : \mathrm{el}(A))$, then p is equal to a substitution of the form $(\tau, y \leftarrow a)$, according to the computation

$$\mathrm{p} = \mathrm{id} \circ \mathrm{p} = (\mathrm{p}, y \leftarrow y) \circ \mathrm{p} = (\mathrm{p} \circ \mathrm{p}, y \leftarrow y \circ \mathrm{p}) : \\ (\Delta, y : \mathrm{el}(A)) \longleftarrow (\Delta, y : \mathrm{el}(A), x : \mathrm{el}(B)).$$

For example,

$$\mathrm{p} = (x \leftarrow x \circ \mathrm{p} \circ \mathrm{p}, y \leftarrow y \circ \mathrm{p}) : \\ (x : \mathrm{el}(A), y : \mathrm{el}(B)) \longleftarrow (x : \mathrm{el}(A), y : \mathrm{el}(B), z : \mathrm{el}(C)).$$

This substitution can be further simplified using inference rules at the very end of this section.

The three remaining forms of substitution are the empty substitution (), the extension substitution $(\sigma, x \leftarrow a)$, and the composition $\sigma \circ \tau$. Of these three, the composition $\sigma \circ \tau$ can be eliminated using the equality (J5.24), if σ is itself a composition, (J5.33), if σ is the empty substitution, and (J5.35), if σ is an extension substitution. In particular, any substitution $\sigma : (\Gamma, x : \mathrm{el}(B)) \longleftarrow \Delta$ is equal, in general in several steps, to a substitution of the form $(\tau, x \leftarrow a)$, where $\tau : \Gamma \longleftarrow \Delta$ and $a : \mathrm{el}(B \circ \tau)\ (\Delta)$.

The following three forms of substitution are frequently used in what follows, and are therefore good to recognize:

$$(\mathrm{id}, x \leftarrow b), \qquad (\sigma \circ \mathrm{p}, y \leftarrow x), \qquad \text{and } (\mathrm{p}, y \leftarrow x).$$

In the latter two forms of substitution, x and y need not be distinct. The first form corresponds to the standard way of substituting a value for a single variable, i.e., $c \circ (\text{id}, x \leftarrow b)$ would be written $c(b/x)$ using ordinary logical notation; the second form corresponds to a *lifted substitution*, as explained below; and the third form corresponds to a change of variable. The substitution $(\text{id}, x \leftarrow b)$ is formed according to the mediate inference rule

$$\frac{b : \text{el}(B)\ (\Gamma)}{(\text{id}, x \leftarrow b) : (\Gamma, x : \text{el}(B)) \longleftarrow \Gamma} \qquad \text{(M5.14)}$$

with the schematic demonstration

$$\cfrac{\cfrac{\Gamma : \text{context}}{\text{id} : \Gamma \longleftarrow \Gamma}\,(\text{R5.2}) \qquad \cfrac{b : \text{el}(B)\ (\Gamma) \quad \cfrac{\cfrac{B : \text{set}\ (\Gamma)}{B \circ \text{id} = B : \text{set}\ (\Gamma)}\,(\text{J5.29})}{B = B \circ \text{id} : \text{set}\ (\Gamma)}\,(\text{M5.2})}{b : \text{el}(B \circ \text{id})\ (\Gamma)}\,(\text{J5.7})}{(\text{id}, x \leftarrow b) : (\Gamma, x : \text{el}(B)) \longleftarrow \Gamma}\,(\text{R5.6}).$$

The substitution $(\sigma \circ \text{p}, y \leftarrow x)$ is formed according to the mediate inference rule

$$\frac{\sigma : \Gamma \longleftarrow \Delta \qquad B : \text{set}\ (\Gamma)}{(\sigma \circ \text{p}, y \leftarrow x) : (\Gamma, y : \text{el}(B)) \longleftarrow (\Delta, x : \text{el}(B \circ \sigma))}\,. \qquad \text{(M5.15)}$$

The schematic demonstration of this inference rule is found on a separate page (p. 130).

The inference rule

$$\frac{B : \text{set}\ (\Gamma)}{(\text{p}, y \leftarrow x) : (\Gamma, y : \text{el}(B)) \longleftarrow (\Gamma, x : \text{el}(B))} \qquad \text{(M5.16)}$$

has the schematic demonstration

$$\cfrac{\cfrac{B : \text{set}\ (\Gamma)}{\text{p} : \Gamma \longleftarrow (\Gamma, x : \text{el}(B))}\,(\text{R5.7}) \qquad \cfrac{B : \text{set}\ (\Gamma)}{x : \text{el}(B \circ \text{p})\ (\Gamma, x : \text{el}(B))}\,(\text{R5.8})}{(\text{p}, y \leftarrow x) : (\Gamma, y : \text{el}(B)) \longleftarrow (\Gamma, x : \text{el}(B))}\,(\text{R5.6})$$

Inference rule (M5.16) can also be demonstrated from (M5.15), by taking σ to be id.

In actual practice, it is cumbersome to keep track of the weakening substitutions. For example, it can be demonstrated that

$$y \circ \text{p} : \text{el}(B \circ \text{p} \circ \text{p})\ (x : \text{el}(A), y : \text{el}(B), z : \text{el}(C)).$$

If the variables involved are distinct, it is possible to eliminate a lot of this redundancy. In the inference rules above, there is no condition on the variables, e.g., (M5.16) is valid also when x and y are taken to be the same variable. In the following inference rule, however, x and y have to be distinct variables:

$$\frac{x : \text{el}(B)\ (\Gamma) \quad A : \text{set}\ (\Gamma)}{x : \text{el}(B \circ \text{p})\ (\Gamma, y : \text{el}(A))}\,. \qquad \text{(R5.9)}$$

The necessity of this condition is clear from the computation rule

$$\frac{\text{el}(B \mid \gamma) : x \mid \gamma \Rightarrow c \,\varepsilon\, \text{el}(C)}{\text{el}(B \circ \text{p} \mid (\gamma, y \leftarrow a)) : x \mid (\gamma, y \leftarrow a) \Rightarrow c \,\varepsilon\, \text{el}(C)}\,, \qquad \text{(C5.9)}$$

The schematic demonstration of inference rule (M5.15) :

$$\dfrac{\dfrac{\sigma : \Gamma \longleftarrow \Delta \quad \dfrac{\dfrac{B : \mathrm{set}\ (\Gamma) \quad \sigma : \Gamma \longleftarrow \Delta}{B \circ \sigma : \mathrm{set}\ (\Delta)}\,\text{(R5.3)}}{\mathrm{p} : \Delta \longleftarrow (\Delta, x : \mathrm{el}(B \circ \sigma))}\,\text{(R5.7)}}{\sigma \circ \mathrm{p} : \Gamma \longleftarrow (\Delta, x : \mathrm{el}(B \circ \sigma))}\,\text{(R5.1)} \quad \dfrac{\dfrac{\dfrac{B : \mathrm{set}\ (\Gamma) \quad \sigma : \Gamma \longleftarrow \Delta}{B \circ \sigma : \mathrm{set}\ (\Delta)}\,\text{(R5.3)}}{x : \mathrm{el}((B \circ \sigma) \circ \mathrm{p})\ (\Delta, x : \mathrm{el}(B \circ \sigma))}\,\text{(R5.8)} \quad \dfrac{\begin{array}{l} B : \mathrm{set}\ (\Gamma) \\ \sigma : \Gamma \longleftarrow \Delta \\ \mathrm{p} : \Delta \longleftarrow (\Delta, x : \mathrm{el}(B \circ \sigma)) \end{array}}{\begin{array}{c} (B \circ \sigma) \circ \mathrm{p} = B \circ (\sigma \circ \mathrm{p}) \\ :\ \mathrm{set}\ (\Delta, x : \mathrm{el}(B \circ \sigma)) \end{array}}\,\text{(J5.31)}}{x : \mathrm{el}(B \circ (\sigma \circ \mathrm{p}))\ (\Delta, x : \mathrm{el}(B \circ \sigma))}\,\text{(J5.7)}}{(\sigma \circ \mathrm{p}, y \leftarrow x) : (\Gamma, y : \mathrm{el}(B)) \longleftarrow (\Delta, x : \mathrm{el}(B \circ \sigma))}\,\text{(R5.6)}.$$

which would be in conflict with (C5.8) if x and y were the same variable. Of course, x and $x \circ \mathrm{p}$ are equal, in the sense of the inference rule

$$\frac{x : \mathrm{el}(B)\ (\Gamma) \quad A : \mathrm{set}\ (\Gamma)}{x \circ \mathrm{p} = x : \mathrm{el}(B \circ \mathrm{p})\ (\Gamma, y : \mathrm{el}(A))} \ . \tag{J5.40}$$

Justification. Let the premisses $x : \mathrm{el}(B)\ (\Gamma)$ and $A : \mathrm{set}\ (\Gamma)$ be given; let γ be an arbitrarily given canonical assignment for Γ and let a be an arbitrarily given canonical element of the set that is the value of $A\,|\,\gamma$, so that $(\gamma, x \leftarrow a)$ is an arbitrary canonical assignment for $(\Gamma, x : \mathrm{el}(A))$. Upon comparing the two computation traces

$$\frac{\mathrm{p}\,|(\gamma, y \leftarrow a) \Rightarrow \gamma\ \varepsilon\ \mathrm{ass}(\Gamma) \quad \mathrm{el}(B\,|\,\gamma) : x\,|\,\gamma \Rightarrow c\ \varepsilon\ \mathrm{el}(C)}{\mathrm{el}(B \circ \mathrm{p}\,|(\gamma, y \leftarrow a)) : x \circ \mathrm{p}\,|(\gamma, y \leftarrow a) \Rightarrow c\ \varepsilon\ \mathrm{el}(C)}$$

and

$$\frac{\mathrm{el}(B\,|\,\gamma) : x\,|\,\gamma \Rightarrow c\ \varepsilon\ \mathrm{el}(C)}{\mathrm{el}(B \circ \mathrm{p}\,|(\gamma, y \leftarrow a)) : x\,|(\gamma, y \leftarrow a) \Rightarrow c\ \varepsilon\ \mathrm{el}(C)} \ ,$$

it becomes clear that $x \circ \mathrm{p}$ and x are equal elements of the set $B \circ \mathrm{p}$ in the context $(\Gamma, x : \mathrm{el}(A))$.

Thus, the example above now becomes

$$y : \mathrm{el}(B \circ \mathrm{p} \circ \mathrm{p})\ (x : \mathrm{el}(A), y : \mathrm{el}(B), z : \mathrm{el}(C)).$$

Still, the weakening substitutions remain on the set B, but only superficially so. Using the inference rules of the next section, it can be demonstrated that

$$y : \mathrm{el}(\mathrm{L}(\mathrm{B}, x))\ (x : \mathrm{el}(\mathrm{N}), y : \mathrm{el}(\mathrm{L}(\mathrm{B}, x)), z : \mathrm{el}(C)),$$

i.e., without mentioning any weakening substitutions.

§ 4. Sets and elements in hypothetical assertions

In this section, the canonical sets introduced in Chapter III, Section 5, are generalized to the noncanonical hypothetical case. Since this generalization follows a simple pattern, only the natural numbers will be treated of in detail.

First, the inference rule

$$\frac{(\Gamma : \mathrm{context})}{\mathrm{N} : \mathrm{set}\ (\Gamma)} \tag{R5.10}$$

is recognized as valid in virtue of the computation rule

$$\mathrm{N}\,|\,\gamma \Rightarrow \mathrm{N}\ \varepsilon\ \mathrm{set}. \tag{C5.10}$$

This computation rule exemplifies the general pattern for a nullary form (categorem) c, i.e., that the value of $c\,|\,\gamma$ is the corresponding canonical nullary form, typically also written c. From any such computation rule, it is easy to justify an inference rule of the form

$$\frac{\sigma : \Gamma \longleftarrow \Delta}{\mathrm{N} \circ \sigma = \mathrm{N} : \mathrm{set}\ (\Delta)} \ . \tag{J5.41}$$

Justification. Let the premiss $\sigma : \Gamma \longleftarrow \Delta$ be given and let δ be an arbitrarily given canonical assignment for Δ. The computation trace for the left-hand side is

$$\mathrm{N}\,|\,\delta \Rightarrow \mathrm{N}\ \varepsilon\ \mathrm{set},$$

and that of the right-hand side is

$$\frac{\sigma\,|\,\delta \Rightarrow \gamma\ \varepsilon\ \mathrm{ass}(\Gamma) \quad \mathrm{N}\,|\,\gamma \Rightarrow \mathrm{N}\ \varepsilon\ \mathrm{set}}{\mathrm{N} \circ \sigma\,|\,\delta \Rightarrow \mathrm{N}\ \varepsilon\ \mathrm{set}} \ .$$

Since the result is N in both cases, the two sides are equal.

Similarly, the inference rule

$$\frac{(\Gamma : \text{context})}{0 : \text{el}(\text{N})\ (\Gamma)} \tag{R5.11}$$

is recognized as valid in virtue of the computation rule

$$\text{el}(\text{N} \mid \gamma) : 0 \mid \gamma \Rightarrow 0\ \varepsilon\ \text{el}(\text{N}). \tag{C5.11}$$

Note that for this computation rule to be well-formed, it is presupposed that $\text{N} \mid \gamma \Rightarrow \text{N}\ \varepsilon$ set. Again, it is trivial to justify the inference rule

$$\frac{\sigma : \Gamma \longleftarrow \Delta}{0 \circ \sigma = 0 : \text{el}(\text{N})\ (\Delta)}\ . \tag{J5.42}$$

The successor operation provides an example of a unary form. The inference rule

$$\frac{n : \text{el}(\text{N})\ (\Gamma)}{\text{s}(n) : \text{el}(\text{N})\ (\Gamma)} \tag{R5.12}$$

is recognized as valid in virtue of the computation rule

$$\frac{\text{el}(\text{N} \mid \gamma) : n \mid \gamma \Rightarrow m\ \varepsilon\ \text{el}(\text{N})}{\text{el}(\text{N} \mid \gamma) : \text{s}(n) \mid \gamma \Rightarrow \text{s}(m)\ \varepsilon\ \text{el}(\text{N})}\ , \tag{C5.12}$$

i.e., $\text{s}(n) \mid \gamma$ is computed eagerly. For forms of higher arity, there are two additional inference rules to establish: first, that two objects of that form are equal if their parts are equal,

$$\frac{n = m : \text{el}(\text{N})\ (\Gamma)}{\text{s}(n) = \text{s}(m) : \text{el}(\text{N})\ (\Gamma)}\ , \tag{J5.43}$$

and next, that substitutions can be moved to the parts of the form,

$$\frac{n : \text{el}(\text{N})\ (\Gamma) \quad \sigma : \Gamma \longleftarrow \Delta}{\text{s}(n) \circ \sigma = \text{s}(n \circ \sigma) : \text{el}(\text{N})\ (\Delta)}\ . \tag{J5.44}$$

Note that the latter inference rule is well-formed due to (J5.41) and (J5.7).

Justification of (J5.43). Let the premiss $n = m : \text{el}(\text{N})\ (\Gamma)$ be given and let γ be an arbitrarily given canonical assignment for Γ. Let $\text{el}(\text{N} \mid \gamma) : \text{s}(n) \mid \gamma \Rightarrow \text{s}(n_0)\ \varepsilon\ \text{el}(\text{N})$, where $\text{el}(\text{N} \mid \gamma) : n \mid \gamma \Rightarrow n_0\ \varepsilon\ \text{el}(\text{N})$, and $\text{el}(\text{N} \mid \gamma) : \text{s}(m) \mid \gamma \Rightarrow \text{s}(m_0)\ \varepsilon\ \text{el}(\text{N})$, where $\text{el}(\text{N} \mid \gamma) : m \mid \gamma \Rightarrow m_0\ \varepsilon\ \text{el}(\text{N})$. That inference rule (J5.43) is valid now means that $\text{s}(n_0)$ and $\text{s}(m_0)$ are equal canonical numbers.

$$\cfrac{\cfrac{n = m : \text{el}(\text{N})\ (\Gamma) \quad \gamma\ \varepsilon\ \text{ass}(\Gamma)}{n \mid \gamma = m \mid \gamma : \text{el}(\text{N} \mid \gamma)}\ (\text{D5.14}) \quad \begin{matrix}\text{el}(\text{N} \mid \gamma) : n \mid \gamma \Rightarrow n_0\ \varepsilon\ \text{el}(\text{N}) \\ \text{el}(\text{N} \mid \gamma) : m \mid \gamma \Rightarrow m_0\ \varepsilon\ \text{el}(\text{N})\end{matrix}}{\cfrac{n_0 = m_0\ \varepsilon\ \text{el}(\text{N})}{\text{s}(n_0) = \text{s}(m_0)\ \varepsilon\ \text{el}(\text{N})}\ (\text{D3.11}).}\ (\text{D4.2})$$

This completes the justification.

Justification of (J5.44). Let the premisses $n : \text{el}(\text{N})\ (\Gamma)$ and $\sigma : \Gamma \longleftarrow \Delta$ be given, and let δ be an arbitrarily given canonical assignment for Δ. Compare the computation traces

$$\cfrac{\cfrac{\sigma \mid \delta \Rightarrow \gamma\ \varepsilon\ \text{ass}(\Gamma) \quad \cfrac{\text{el}(\text{N} \mid \gamma) : n \mid \gamma \Rightarrow m\ \varepsilon\ \text{el}(\text{N})}{\text{el}(\text{N} \mid \gamma) : \text{s}(n) \mid \gamma \Rightarrow \text{s}(m)\ \varepsilon\ \text{el}(\text{N})}}{\text{el}(\text{N} \circ \sigma \mid \delta) : \text{s}(n) \circ \sigma \mid \delta \Rightarrow \text{s}(m)\ \varepsilon\ \text{el}(\text{N})} \quad \text{N} = \text{N} \circ \sigma : \text{set}}{\text{el}(\text{N} \mid \delta) : \text{s}(n) \circ \sigma \mid \delta \Rightarrow \text{s}(m)\ \varepsilon\ \text{el}(\text{N})}$$

and

$$\dfrac{\dfrac{\sigma\,|\,\delta \Rightarrow \gamma \;\varepsilon\; \mathrm{ass}(\Gamma) \quad \mathrm{el}(\mathrm{N}\,|\,\gamma) : n\,|\,\gamma \Rightarrow m \;\varepsilon\; \mathrm{el}(\mathrm{N})}{\mathrm{el}(\mathrm{N}\circ\sigma\,|\,\delta) : n\circ\sigma\,|\,\delta \Rightarrow m \;\varepsilon\; \mathrm{el}(\mathrm{N})} \qquad \mathrm{N} = \mathrm{N}\circ\sigma : \mathrm{set}}{\dfrac{\mathrm{el}(\mathrm{N}\,|\,\delta) : n\circ\sigma\,|\,\delta \Rightarrow m \;\varepsilon\; \mathrm{el}(\mathrm{N})}{\mathrm{el}(\mathrm{N}\,|\,\delta) : \mathrm{s}(n\circ\sigma)\,|\,\delta \Rightarrow \mathrm{s}(m) \;\varepsilon\; \mathrm{el}(\mathrm{N})}}\ .$$

They show that the two sides are equal.

For the other sets and set-forming operations, no explanations or justifications will be given—only the inference rules.

The set of Booleans.

$$\frac{(\Gamma : \mathrm{context})}{\mathrm{B} : \mathrm{set}\ (\Gamma)}, \quad (\mathrm{R}5.13) \qquad \mathrm{B}\,|\,\gamma \Rightarrow \mathrm{B} \;\varepsilon\; \mathrm{set}, \quad (\mathrm{C}5.13)$$

$$\frac{\sigma : \Gamma \longleftarrow \Delta}{\mathrm{B}\circ\sigma = \mathrm{B} : \mathrm{set}\ (\Delta)}; \quad (\mathrm{J}5.45) \qquad \frac{(\Gamma : \mathrm{context})}{1 : \mathrm{el}(\mathrm{B})\ (\Gamma)}, \quad (\mathrm{R}5.14)$$

$$\mathrm{el}(\mathrm{B}\,|\,\gamma) : 1\,|\,\gamma \Rightarrow 1 \;\varepsilon\; \mathrm{el}(\mathrm{B}), \quad (\mathrm{C}5.14) \qquad \frac{\sigma : \Gamma \longleftarrow \Delta}{1\circ\sigma = 1 : \mathrm{el}(\mathrm{B})\ (\Delta)}; \quad (\mathrm{J}5.46)$$

$$\frac{(\Gamma : \mathrm{context})}{0 : \mathrm{el}(\mathrm{B})\ (\Gamma)}, \quad (\mathrm{R}5.15) \qquad \mathrm{el}(\mathrm{B}\,|\,\gamma) : 0\,|\,\gamma \Rightarrow 0 \;\varepsilon\; \mathrm{el}(\mathrm{B}), \quad (\mathrm{C}5.15)$$

$$\frac{\sigma : \Gamma \longleftarrow \Delta}{0\circ\sigma = 0 : \mathrm{el}(\mathrm{B})\ (\Delta)}\ . \quad (\mathrm{J}5.47)$$

The unit set.

$$\frac{(\Gamma : \mathrm{set})}{1 : \mathrm{set}\ (\Gamma)}, \quad (\mathrm{R}5.16) \qquad 1\,|\,\gamma \Rightarrow 1 \;\varepsilon\; \mathrm{set}, \quad (\mathrm{C}5.16)$$

$$\frac{\sigma : \Gamma \longleftarrow \Delta}{1\circ\sigma = 1 : \mathrm{set}\ (\Delta)}; \qquad \frac{(\Gamma : \mathrm{context})}{0 : \mathrm{el}(1)\ (\Gamma)}, \quad (\mathrm{R}5.17)$$

$$\mathrm{el}(1\,|\,\gamma) : 0\,|\,\gamma \Rightarrow 0 \;\varepsilon\; \mathrm{el}(1), \quad (\mathrm{C}5.17) \qquad \frac{\sigma : \Gamma \longleftarrow \Delta}{0\circ\sigma = 0 : \mathrm{el}(1)\ (\Delta)}\ . \quad (\mathrm{J}5.48)$$

The empty set.

$$\frac{(\Gamma : \mathrm{set})}{\emptyset : \mathrm{set}\ (\Gamma)}, \quad (\mathrm{R}5.18) \qquad \emptyset\,|\,\gamma \Rightarrow \emptyset \;\varepsilon\; \mathrm{set}, \quad (\mathrm{C}5.18)$$

$$\frac{\sigma : \Gamma \longleftarrow \Delta}{\emptyset\circ\sigma = \emptyset : \mathrm{set}\ (\Delta)}\ . \quad (\mathrm{J}5.49)$$

The Cartesian product.

$$\frac{A : \mathrm{set}\ (\Gamma) \quad B : \mathrm{set}\ (\Gamma)}{A \times B : \mathrm{set}\ (\Gamma)}, \quad (\mathrm{R}5.19)$$

$$\frac{A\,|\,\gamma \Rightarrow C \;\varepsilon\; \mathrm{set} \quad B\,|\,\gamma \Rightarrow D \;\varepsilon\; \mathrm{set}}{A \times B\,|\,\gamma \Rightarrow C \times D \;\varepsilon\; \mathrm{set}}, \quad (\mathrm{C}5.19)$$

$$\frac{A = C : \mathrm{set}\ (\Gamma) \quad B = D : \mathrm{set}\ (\Gamma)}{A \times B = C \times D : \mathrm{set}\ (\Gamma)}\ , \tag{J5.50}$$

$$\frac{A : \mathrm{set}\ (\Gamma) \quad B : \mathrm{set}\ (\Gamma) \quad \sigma : \Gamma \longleftarrow \Delta}{(A \times B) \circ \sigma = (A \circ \sigma) \times (B \circ \sigma) : \mathrm{set}\ (\Delta)}\ ; \tag{J5.51}$$

$$\frac{a : \mathrm{el}(A)\ (\Gamma) \quad b : \mathrm{el}(B)\ (\Gamma)}{(a\,, b) : \mathrm{el}(A \times B)\ (\Gamma)}\ , \tag{R5.20}$$

$$\frac{\mathrm{el}(A \,|\, \gamma) : a \,|\, \gamma \Rightarrow c\ \varepsilon\ \mathrm{el}(C) \quad \mathrm{el}(B \,|\, \gamma) : b \,|\, \gamma \Rightarrow d\ \varepsilon\ \mathrm{el}(D)}{\mathrm{el}(A \times B \,|\, \gamma) : (a\,, b) \,|\, \gamma \Rightarrow (c, d)\ \varepsilon\ \mathrm{el}(C \times D)}\ , \tag{C5.20}$$

$$\frac{a = c : \mathrm{el}(A)\ (\Gamma) \quad b = d : \mathrm{el}(B)\ (\Gamma)}{(a\,, b) = (c\,, d) : \mathrm{el}(A \times B)\ (\Gamma)}\ , \tag{J5.52}$$

$$\frac{a : \mathrm{el}(A)\ (\Gamma) \quad b : \mathrm{el}(B)\ (\Gamma) \quad \sigma : \Gamma \longleftarrow \Delta}{(a\,, b) \circ \sigma = (a \circ \sigma\,, b \circ \sigma) : \mathrm{el}((A \times B) \circ \sigma)\ (\Delta)}\ . \tag{J5.53}$$

The disjoint union.

$$\frac{A : \mathrm{set}\ (\Gamma) \quad B : \mathrm{set}\ (\Gamma)}{A + B : \mathrm{set}\ (\Gamma)}\ , \tag{R5.21}$$

$$\frac{A \,|\, \gamma \Rightarrow C\ \varepsilon\ \mathrm{set} \quad B \,|\, \gamma \Rightarrow D\ \varepsilon\ \mathrm{set}}{A + B \,|\, \gamma \Rightarrow C + D\ \varepsilon\ \mathrm{set}}\ , \tag{C5.21}$$

$$\frac{A = C : \mathrm{set}\ (\Gamma) \quad B = D : \mathrm{set}\ (\Gamma)}{A + B = C + D : \mathrm{set}\ (\Gamma)}\ , \tag{J5.54}$$

$$\frac{A : \mathrm{set}\ (\Gamma) \quad B : \mathrm{set}\ (\Gamma) \quad \sigma : \Gamma \longleftarrow \Delta}{(A + B) \circ \sigma = (A \circ \sigma) + (B \circ \sigma) : \mathrm{set}\ (\Delta)}\ ; \tag{J5.55}$$

$$\frac{a : \mathrm{el}(A)\ (\Gamma) \quad (B : \mathrm{set}\ (\Gamma))}{\mathrm{i}(a) : \mathrm{el}(A + B)\ (\Gamma)}\ , \tag{R5.22}$$

$$\frac{\mathrm{el}(A \,|\, \gamma) : a \,|\, \gamma \Rightarrow c\ \varepsilon\ \mathrm{el}(C) \quad (B \,|\, \gamma \Rightarrow D\ \varepsilon\ \mathrm{set})}{\mathrm{el}(A + B \,|\, \gamma) : \mathrm{i}(a) \,|\, \gamma \Rightarrow \mathrm{i}(c)\ \varepsilon\ \mathrm{el}(C + D)}\ , \tag{C5.22}$$

$$\frac{a = c : \mathrm{el}(A)\ (\Gamma) \quad (B : \mathrm{set}\ (\Gamma))}{\mathrm{i}(a) = \mathrm{i}(c) : \mathrm{el}(A + B)\ (\Gamma)}\ , \tag{J5.56}$$

$$\frac{a : \mathrm{el}(A)\ (\Gamma) \quad B : \mathrm{set}\ (\Gamma) \quad \sigma : \Gamma \longleftarrow \Delta}{\mathrm{i}(a) \circ \sigma = \mathrm{i}(a \circ \sigma) : \mathrm{el}((A + B) \circ \sigma)\ (\Delta)}\ ; \tag{J5.57}$$

$$\frac{(A : \mathrm{set}\ (\Gamma)) \quad b : \mathrm{el}(B)\ (\Gamma)}{\mathrm{j}(b) : \mathrm{el}(A + B)\ (\Gamma)}\ , \tag{R5.23}$$

$$\frac{(A \,|\, \gamma \Rightarrow C\ \varepsilon\ \mathrm{set}) \quad \mathrm{el}(B \,|\, \gamma) : b \,|\, \gamma \Rightarrow d\ \varepsilon\ \mathrm{el}(D)}{\mathrm{el}(A + B \,|\, \gamma) : \mathrm{j}(b) \,|\, \gamma \Rightarrow \mathrm{j}(d)\ \varepsilon\ \mathrm{el}(C + D)}\ , \tag{C5.23}$$

$$\frac{(A : \mathrm{set}\ (\Gamma)) \quad b = d : \mathrm{el}(B)\ (\Gamma)}{\mathrm{j}(b) = \mathrm{j}(d) : \mathrm{el}(A + B)\ (\Gamma)}\ , \tag{J5.58}$$

$$\frac{A : \mathrm{set}\ (\Gamma) \quad b : \mathrm{el}(B)\ (\Gamma) \quad \sigma : \Gamma \longleftarrow \Delta}{\mathrm{j}(b) \circ \sigma = \mathrm{j}(b \circ \sigma) : \mathrm{el}((A + B) \circ \sigma)\ (\Delta)}\ . \tag{J5.59}$$

The set of lists.

$$\frac{A : \mathrm{set}\ (\Gamma) \quad m : \mathrm{el}(\mathrm{N})\ (\Gamma)}{\mathrm{L}(A, m) : \mathrm{set}\ (\Gamma)}\ , \tag{R5.24}$$

$$\frac{A\,|\,\gamma \Rightarrow B \;\varepsilon\; \text{set} \quad \text{el}(\text{N}\,|\,\gamma) : m\,|\,\gamma \Rightarrow n \;\varepsilon\; \text{el}(\text{N})}{\text{L}(A,m)\,|\,\gamma \Rightarrow \text{L}(B,n) \;\varepsilon\; \text{set}}\,, \tag{C5.24}$$

$$\frac{A = B : \text{set}\ (\Gamma) \quad m = n : \text{el}(\text{N})\ (\Gamma)}{\text{L}(A,m) = \text{L}(B,n) : \text{set}\ (\Gamma)}\,, \tag{J5.60}$$

$$\frac{A : \text{set}\ (\Gamma) \quad m : \text{el}(\text{N})\ (\Gamma) \quad \sigma : \Gamma \longleftarrow \Delta}{\text{L}(A,m) \circ \sigma = \text{L}(A \circ \sigma, m \circ \sigma) : \text{set}\ (\Delta)}\,; \tag{J5.61}$$

$$\frac{(A : \text{set}\ (\Gamma))}{() : \text{el}(\text{L}(A,0))\ (\Gamma)}\,, \tag{R5.25}$$

$$\frac{(A\,|\,\gamma \Rightarrow B \;\varepsilon\; \text{set})}{\text{el}(\text{L}(A,0)\,|\,\gamma) : ()\,|\,\gamma \Rightarrow () \;\varepsilon\; \text{el}(\text{L}(B,0))}\,, \tag{C5.25}$$

$$\frac{A : \text{set}\ (\Gamma) \quad \sigma : \Gamma \longleftarrow \Delta}{() \circ \sigma = () : \text{el}(\text{L}(A,0) \circ \sigma)\ (\Delta)}\,; \tag{J5.62}$$

$$\frac{a : \text{el}(A)\ (\Gamma) \quad l : \text{el}(\text{L}(A,p))\ (\Gamma)}{(a\,,l) : \text{el}(\text{L}(A,\text{s}(p)))\ (\Gamma)}\,, \tag{R5.26}$$

$$\frac{\text{el}(A\,|\,\gamma) : a\,|\,\gamma \Rightarrow b \;\varepsilon\; \text{el}(B) \quad \text{el}(\text{L}(A,p)\,|\,\gamma) : l\,|\,\gamma \Rightarrow m \;\varepsilon\; \text{el}(\text{L}(B,q))}{\text{el}(\text{L}(A,\text{s}(p))\,|\,\gamma) : (a\,,l)\,|\,\gamma \Rightarrow (b\,,m) \;\varepsilon\; \text{el}(\text{L}(B,\text{s}(q)))}\,, \tag{C5.26}$$

$$\frac{a = b : \text{el}(A)\ (\Gamma) \quad l = m : \text{el}(\text{L}(A,p))\ (\Gamma)}{(a\,,l) = (b\,,m) : \text{el}(\text{L}(A,\text{s}(p)))\ (\Gamma)}\,, \tag{J5.63}$$

$$\frac{a : \text{el}(A)\ (\Gamma) \quad l : \text{el}(\text{L}(A,p))\ (\Gamma) \quad \sigma : \Gamma \longleftarrow \Delta}{(a\,,l) \circ \sigma = (a \circ \sigma\,, l \circ \sigma) : \text{el}(\text{L}(A,\text{s}(p)) \circ \sigma)\ (\Delta)}\,. \tag{J5.64}$$

§ 5. Closures and the λ-calculus

The notion of a *closure* is familiar from computer science.[11] As a first approximation, a closure can be described as an open expression

$$f(x_1, \ldots, x_n, x_{n+1})$$

together with an assignment of values

$$(x_1 \leftarrow a_1, \ldots, x_n \leftarrow a_n)$$

to the first variables. The pair of the expression and the assignment is then viewed as the function in the modern sense which takes a to $f(a_1, \ldots, a_n, a)$. Interestingly, a detailed scrutiny of the meaning explanations of intuitionistic type theory reveals that the canonical objects of the logical category fam(C) and the canonical elements of the set $\Pi(C, F)$ are closures.[12]

First I will consider the logical category fam(C), where C is a canonical set. The above explanation of closures as canonical objects of this logical category is made explicit by the inference rule

$$\frac{B : \text{set}\ (\Gamma, x : \text{el}(A)) \quad \gamma \;\varepsilon\; \text{ass}(\Gamma) \quad A\,|\,\gamma \Rightarrow C \;\varepsilon\; \text{set}}{\text{cl}(\hat{x}B, \gamma) \;\varepsilon\; \text{fam}(C)}\,. \tag{R5.27}$$

[11] Cf. Landin, 'The mechanical evaluation of expressions', p. 318.

[12] The logical category fam(C) and the canonical set $\Pi(C, F)$ were introduced in Ch. IV, § 7.

The notation $\hat{x}B$, due to Russell,[13] is to indicate that x becomes bound in B. An alternative notation is $(x)B$, but, to my mind, parentheses are already sufficiently overloaded in mathematical notation. Recall that $F \;\varepsilon\; \text{fam}(C)$ means that, if $c \;\varepsilon\; \text{el}(C)$, then $\text{app}[F, c]$: set and that, if $c = d \;\varepsilon\; \text{el}(C)$, then $\text{app}[F, c] = \text{app}[F, d]$: set. The previous inference rule is associated with the computation rule

$$\frac{B\,|(\gamma, x \leftarrow c) \Rightarrow D \;\varepsilon\; \text{set}}{\text{app}[\text{cl}(\hat{x}B, \gamma), c] \Rightarrow D \;\varepsilon\; \text{set}}\,, \tag{C5.27}$$

in which c is an arbitrarily given canonical element of the canonical set C. In view of (D5.11) and (D5.6), it is clear that, if $c = d \;\varepsilon\; \text{el}(C)$, then $\text{app}[\text{cl}(\hat{x}B, \gamma), c]$; and $\text{app}[\text{cl}(\hat{x}B, \gamma), d]$ are equal sets. Therefore the inference rule (R5.27) is valid. From the above computation rule, the inference rule

$$\frac{B : \text{set}\ (\Gamma, x : \text{el}(A)) \quad \gamma \;\varepsilon\; \text{ass}(\Gamma) \quad (A\,|\,\gamma \Rightarrow C \;\varepsilon\; \text{set}) \quad c \;\varepsilon\; \text{el}(C)}{\text{app}[\text{cl}(\hat{x}B, \gamma), c] = B\,|(\gamma, x \leftarrow c) : \text{set}} \tag{J5.65}$$

is trivial to justify. Moreover, two closures are equal if their parts are equal, i.e.,

$$\frac{B = C : \text{set}\ (\Gamma, x : \text{el}(A)) \quad \gamma = \delta \;\varepsilon\; \text{ass}(\Gamma) \quad (A\,|\,\delta \Rightarrow D \;\varepsilon\; \text{set})}{\text{cl}(\hat{x}B, \gamma) = \text{cl}(\hat{x}C, \delta) \;\varepsilon\; \text{fam}(D)}\,. \tag{J5.66}$$

Note, however, that two closures may be equal even if their parts are not equal, as seen in the justifications of (J5.68) and (η).

Justification. Let the premisses be given and let d be an arbitrarily given canonical element of D. The desired conclusion follows from the demonstration

$$\cfrac{B = C : \text{set}\ (\Gamma, x : \text{el}(A)) \qquad \cfrac{\gamma = \delta \;\varepsilon\; \text{ass}(\Gamma) \quad (A\,|\,\delta \Rightarrow D \;\varepsilon\; \text{set}) \quad \cfrac{d \;\varepsilon\; \text{el}(D)}{d = d \;\varepsilon\; \text{el}(D)}\,\text{(D3.1)}}{(\gamma, x \leftarrow d) = (\delta, x \leftarrow d) \;\varepsilon\; \text{ass}(\Gamma, x : \text{el}(A))}\,\text{(D5.11)}}{B\,|(\gamma, x \leftarrow d) = C\,|(\delta, x \leftarrow d) : \text{set}}\,\text{(M5.1)}$$

together with (J5.65) and the fact that the equality relation between noncanonical sets is an equivalence relation.

The inference rule

$$\frac{A : \text{set}\ (\Gamma) \quad B : \text{set}\ (\Gamma, x : \text{el}(A))}{(\Pi\, x : A)B : \text{set}\ (\Gamma)} \tag{R5.28}$$

is now recognized as valid in virtue of the computation rule

$$\frac{A\,|\,\gamma \Rightarrow C \;\varepsilon\; \text{set}}{(\Pi\, x : A)B\,|\,\gamma \Rightarrow \Pi(C, \text{cl}(\hat{x}B, \gamma)) \;\varepsilon\; \text{set}}\,. \tag{C5.28}$$

[13]Frege used the notation $\grave{x}$, i.e., *spiritus lenis*, in his *Grundgesetze der Arithmetik I*, §9, and Russell used $\hat{x}$, i.e., circumflex, in 'Mathematical Logic as Based on the Theory of Types', p. 250, with a similar meaning. Here (like in Russell, loc. cit.), $\hat{x}B$ is a "fictitious object" ; moreover, the notation $\hat{x}$ is not a part of the analysis of expressions as syntax trees (p. 7) : in line with de Bruijn's suggestion (*Recommendations concerning standardization of mathematical formulas*), the form of expression $\text{cl}(\hat{x}B, \gamma)$ is analyzed as having the form cl, with two parts B and γ, and the line connecting cl with B in the syntax tree is labelled with an x. I will also use the convention that the scope of $\hat{x}$ extends as far to the right as possible. The same applies to all variable binding operations (Π, λ, Σ, R, etc.).

Note that $(\Pi x : A)B$ is a variable binding operation that binds x in B; a more uniform notation would be $\Pi(A, \hat{x}B)$.

$$\frac{A = C : \mathrm{set}\ (\Gamma) \quad B = D : \mathrm{set}\ (\Gamma, x : \mathrm{el}(A))}{(\Pi x : A)B = (\Pi x : C)D : \mathrm{set}\ (\Gamma)} . \tag{J5.67}$$

Justification. Let the premisses be given and let γ be an arbitrarily given canonical assignment for the context Γ. Let further $A \,|\, \gamma \Rightarrow A_0 \ \varepsilon\ \mathrm{set}$ and $C \,|\, \gamma \Rightarrow C_0 \ \varepsilon\ \mathrm{set}$. It remains to show that $\Pi(A_0, \mathrm{cl}(\hat{x}B, \gamma))$ and $\Pi(C_0, \mathrm{cl}(\hat{x}D, \gamma))$ are equal canonical sets. According to (D5.7) and (D4.1), $A_0 = C_0 \ \varepsilon\ \mathrm{set}$; furthermore, (J5.3) and (J5.66) show that $\mathrm{cl}(\hat{x}B, \gamma)$ and $\mathrm{cl}(\hat{x}D, \gamma)$ are equal families of sets over C_0. The conclusion now follows from (J4.15).

It also has to be explained how the form of expression $(\Pi x : A)B$ works together with substitutions:

$$\frac{A : \mathrm{set}\ (\Gamma) \quad B : \mathrm{set}\ (\Gamma, x : \mathrm{el}(A)) \quad \sigma : \Gamma \longleftarrow \Delta}{((\Pi x : A)B) \circ \sigma = (\Pi y : A \circ \sigma)(B \circ (\sigma \circ \mathrm{p}, x \leftarrow y)) : \mathrm{set}\ (\Delta)} . \tag{J5.68}$$

Note that if σ is the identity substitution, this inference rule gives change of bound variable as a special case.

Justification. Let the premisses be given and let δ be an arbitrarily given assignment for the context Δ. Now compare the computation traces

$$\dfrac{\sigma \,|\, \delta \Rightarrow \gamma \ \varepsilon\ \mathrm{ass}(\Gamma) \quad \dfrac{A \,|\, \gamma \Rightarrow C \ \varepsilon\ \mathrm{set}}{(\Pi x : A)B \,|\, \gamma \Rightarrow \Pi(C, \mathrm{cl}(\hat{x}B, \gamma)) \ \varepsilon\ \mathrm{set}}}{((\Pi x : A)B) \circ \sigma \,|\, \delta \Rightarrow \Pi(C, \mathrm{cl}(\hat{x}B, \gamma)) \ \varepsilon\ \mathrm{set}}$$

and

$$\dfrac{\dfrac{\sigma \,|\, \delta \Rightarrow \gamma \ \varepsilon\ \mathrm{ass}(\Delta) \quad A \,|\, \gamma \Rightarrow C \ \varepsilon\ \mathrm{set}}{A \circ \sigma \,|\, \delta \Rightarrow C \ \varepsilon\ \mathrm{set}}}{(\Pi y : A \circ \sigma)(B \circ (\sigma \circ \mathrm{p}, x \leftarrow y)) \,|\, \delta \Rightarrow \Pi(C, \mathrm{cl}(\hat{y}B \circ (\sigma \circ \mathrm{p}, x \leftarrow y), \delta)) \ \varepsilon\ \mathrm{set}} .$$

It remains to show that their values are equal canonical sets. In view of (J4.15), it suffices to show that $\mathrm{cl}(\hat{x}B, \gamma)$ and $\mathrm{cl}(\hat{y}B \circ (\sigma \circ \mathrm{p}, x \leftarrow y), \delta)$ are equal families of sets over the canonical set C. Thus, let c be an arbitrarily given canonical element of C, and compare the computation traces

$$\dfrac{B \,|\, (\gamma, x \leftarrow c) \Rightarrow D \ \varepsilon\ \mathrm{set}}{\mathrm{app}[\mathrm{cl}(\hat{x}B, \gamma), c] \Rightarrow D \ \varepsilon\ \mathrm{set}}$$

and

$$\dfrac{\dfrac{\dfrac{\dfrac{\begin{array}{l}\mathrm{p} \,|\, (\delta, y \leftarrow c) \Rightarrow \delta \ \varepsilon\ \mathrm{ass}(\Delta) \\ \sigma \,|\, \delta \Rightarrow \gamma \ \varepsilon\ \mathrm{ass}(\Gamma)\end{array}}{\sigma \circ \mathrm{p} \,|\, (\delta, y \leftarrow c) \Rightarrow \gamma \ \varepsilon\ \mathrm{ass}(\Gamma)} \quad \begin{array}{l}\mathrm{el}(A \,|\, \gamma) : \\ y \,|\, (\delta, y \leftarrow c) \Rightarrow c \ \varepsilon\ \mathrm{el}(C)\end{array}}{(\sigma \circ \mathrm{p}, x \leftarrow y) \,|\, (\delta, y \leftarrow c) \Rightarrow (\gamma, x \leftarrow c) \ \varepsilon\ \mathrm{ass}(\Gamma, x : \mathrm{el}(A))} \quad B \,|\, (\gamma, x \leftarrow c) \Rightarrow D \ \varepsilon\ \mathrm{set}}{B \circ (\sigma \circ \mathrm{p}, x \leftarrow y) \,|\, (\delta, y \leftarrow c) \Rightarrow D \ \varepsilon\ \mathrm{set}}}{\mathrm{app}[\mathrm{cl}(\hat{y}B \circ (\sigma \circ \mathrm{p}, x \leftarrow y), \delta), c] \Rightarrow D \ \varepsilon\ \mathrm{set}}$$

Note that $\sigma \,|\, \delta \Rightarrow \gamma \ \varepsilon\ \mathrm{ass}(\Gamma)$ according to the above. This completes the justification.

In understanding the above justification, it is of benefit to recall that $(\sigma \circ \mathrm{p}, x \leftarrow y)$ is the *lifted substitution* introduced on p. 129.

Closures are canonical elements of the set $\Pi(C, F)$ in a manner similar to how closures are canonical families of sets.

$$\frac{b : \mathrm{el}(B)\ (\Gamma, x : \mathrm{el}(A)) \quad \gamma \ \varepsilon\ \mathrm{ass}(\Gamma) \quad A \,|\, \gamma \Rightarrow C \ \varepsilon\ \mathrm{set}}{\mathrm{cl}(\hat{x}b, \gamma) \ \varepsilon\ \mathrm{el}(\Pi(C, \mathrm{cl}(\hat{x}B, \gamma)))} . \tag{R5.29}$$

In this case, the computation rule is given by

$$\frac{\mathrm{el}(B\,|\,(\gamma, x \leftarrow c)) : b\,|\,(\gamma, x \leftarrow c) \Rightarrow d \;\varepsilon\; \mathrm{el}(D)}{\mathrm{el}(\mathrm{app}[\mathrm{cl}(\hat{x}B, \gamma), c]) : \mathrm{app}[\mathrm{cl}(\hat{x}b, \gamma), c] \Rightarrow d \;\varepsilon\; \mathrm{el}(D)} \,. \tag{C5.29}$$

As for families of sets, it is trivial to justify the inference rule

$$\frac{b : \mathrm{el}(B)\ (\mathit{\Gamma}, x : \mathrm{el}(A)) \quad \gamma \;\varepsilon\; \mathrm{ass}(\mathit{\Gamma}) \quad A\,|\,\gamma \Rightarrow C \;\varepsilon\; \mathrm{set} \quad c \;\varepsilon\; \mathrm{el}(C)}{\mathrm{app}[\mathrm{cl}(\hat{x}b, \gamma), c] = b\,|\,(\gamma, x \leftarrow c) : \mathrm{el}(B\,|\,(\gamma, x \leftarrow c))} \,. \tag{J5.69}$$

Two elements of the form $\mathrm{cl}(\hat{x}b, \gamma)$ are equal if their parts are equal:

$$\frac{b = c : \mathrm{el}(B)\ (\mathit{\Gamma}, x : \mathrm{el}(A)) \quad \gamma = \delta \;\varepsilon\; \mathrm{ass}(\mathit{\Gamma}) \quad A\,|\,\delta \Rightarrow D \;\varepsilon\; \mathrm{set}}{\mathrm{cl}(\hat{x}b, \gamma) = \mathrm{cl}(\hat{x}c, \delta) \;\varepsilon\; \mathrm{el}(\Pi(D, \mathrm{cl}(\hat{x}B, \delta)))} \,. \tag{J5.70}$$

Justification. Let the premisses be given and let d be an arbitrarily given canonical element of D. The desired conclusion follows from the demonstration

$$\dfrac{b = c : \mathrm{el}(B)\ (\mathit{\Gamma}, x : \mathrm{el}(A)) \qquad \dfrac{\gamma = \delta \;\varepsilon\; \mathrm{ass}(\mathit{\Gamma}) \quad (A\,|\,\delta \Rightarrow D \;\varepsilon\; \mathrm{set}) \quad \dfrac{d \;\varepsilon\; \mathrm{el}(D)}{d = d \;\varepsilon\; \mathrm{el}(D)}\,(\mathrm{D3.1})}{(\gamma, x \leftarrow d) = (\delta, x \leftarrow d) \;\varepsilon\; \mathrm{ass}(\mathit{\Gamma}, x : \mathrm{el}(A))}\,(\mathrm{D5.11})}{b\,|\,(\gamma, x \leftarrow d) = c\,|\,(\delta, x \leftarrow d) : \mathrm{el}(B\,|\,(\delta, x \leftarrow d))}\,(\mathrm{M5.6})$$

together with (J5.69) and the fact that the equality relation between noncanonical elements of a set is an equivalence relation.

Closures are the canonical forms of λ abstractions. That is, the usual rule for λ abstraction

$$\frac{b : \mathrm{el}(B)\ (\mathit{\Gamma}, x : \mathrm{el}(A))}{(\lambda\, x)b : \mathrm{el}((\Pi\, x : A)B)\ (\mathit{\Gamma})} \tag{R5.30}$$

is recognized as valid in virtue of the computation rule

$$\frac{(A\,|\,\gamma \Rightarrow C \;\varepsilon\; \mathrm{set})}{\mathrm{el}((\Pi\, x : A)B\,|\,\gamma) : (\lambda\, x)b\,|\,\gamma \Rightarrow \mathrm{cl}(\hat{x}b, \gamma) \;\varepsilon\; \mathrm{el}(\Pi(C, \mathrm{cl}(\hat{x}B, \gamma)))} \,. \tag{C5.30}$$

Note that $(\lambda\, x)b$ is a variable binding operation; a more uniform notation would be $\lambda(\hat{x}b)$.

In λ-calculus, the inference rule that shows that two elements of λ form are equal if their parts are equal, i.e., the inference rule

$$\frac{b = c : \mathrm{el}(B)\ (\mathit{\Gamma}, x : \mathrm{el}(A))}{(\lambda\, x)b = (\lambda\, x)c : \mathrm{el}((\Pi\, x : A)B)\ (\mathit{\Gamma})} \,, \tag{ξ}$$

is called the ξ rule.

Justification of (ξ). Let the premiss be given and let γ be an arbitrarily given canonical assignment for $\mathit{\Gamma}$. It now follows from (J5.3) and (J5.70) that the values of $(\lambda\, x)b\,|\,\gamma$ and $(\lambda\, x)c\,|\,\gamma$, viz., $\mathrm{cl}(\hat{x}b, \gamma)$ and $\mathrm{cl}(\hat{x}c, \gamma)$, respectively, are equal.

It also has to be verified that substitutions can be moved under λ abstractions:

$$\frac{b : \mathrm{el}(B)\ (\mathit{\Gamma}, x : \mathrm{el}(A)) \quad \sigma : \mathit{\Gamma} \longleftarrow \mathit{\Delta}}{((\lambda x)b)\circ\sigma = (\lambda y)(b\circ(\sigma\circ\mathrm{p}, x \leftarrow y)) : \mathrm{el}(((\Pi x : A)B)\circ\sigma)\,(\mathit{\Delta})} \,. \tag{J5.71}$$

Note that this inference rule is well-formed due to (M5.15).

Justification of (J5.71). Let the premisses be given and let δ be an arbitrarily given assignment for the context Δ. Let the value of $\sigma \,|\, \delta$ be $\gamma \;\varepsilon\; \mathrm{ass}(\Gamma)$. The value of the left-hand side of the equality in the conclusion is $\mathrm{cl}(\hat{x}b, \gamma) \;\varepsilon\; \mathrm{el}(\Pi(C, \mathrm{cl}(\hat{x}B, \gamma)))$, where $A \,|\, \gamma \Rightarrow C \;\varepsilon$ set, and the value of the right-hand side is $\mathrm{cl}(\hat{y}b \circ (\sigma \circ \mathrm{p}, x \leftarrow y), \delta) \;\varepsilon\; \mathrm{el}(\Pi(C, \mathrm{cl}(\hat{x}B, \gamma)))$. It remains to show that these two elements are equal. Thus, let c be an arbitrarily given canonical element of C; $\mathrm{el}(\mathrm{app}[\mathrm{cl}(\hat{x}B, \gamma), c]) : \mathrm{app}[\mathrm{cl}(\hat{x}b, \gamma), c] \Rightarrow d \;\varepsilon\; \mathrm{el}(D)$ follows from $\mathrm{el}(B \,|(\gamma, x \leftarrow c)) : b \,|(\gamma, x \leftarrow c) \Rightarrow d \;\varepsilon\; \mathrm{el}(D)$; now consider the computation of $\mathrm{app}[\mathrm{cl}(\hat{y}b \circ (\sigma \circ \mathrm{p}, x \leftarrow y), \delta), c] : \mathrm{app}[\mathrm{cl}(\hat{x}B, \gamma), c]$; the first step is a set conversion, according to (C4.1); the next step consists in a change from application of closures to substitutions, according to (C5.30), so the value of $\mathrm{app}[\mathrm{cl}(\hat{y}b \circ (\sigma \circ \mathrm{p}, x \leftarrow y), \delta), c]$ is the value of $b \circ (\sigma \circ \mathrm{p}, x \leftarrow y) \,|(\delta, y \leftarrow c)$. As seen in the justification of (J5.68), the value of $(\sigma \circ \mathrm{p}, x \leftarrow y) \,|(\delta, y \leftarrow c)$ is $(\gamma, x \leftarrow c)$, where $\sigma \,|\, \delta \Rightarrow \gamma \;\varepsilon\; \mathrm{ass}(\Gamma)$; accordingly, the value of $b \,|(\gamma, x \leftarrow c)$ is $d \;\varepsilon\; \mathrm{el}(D)$. This completes the justification.

The next topic is application of a hypothetical function to a hypothetical argument, given by the inference rule

$$\frac{f : \mathrm{el}((\Pi\, x : A)B)\ (\Gamma) \quad a : \mathrm{el}(A)\ (\Gamma)}{\mathrm{app}(f, a) : \mathrm{el}(B \circ (\mathrm{id}, x \leftarrow a))\ (\Gamma)}, \tag{R5.31}$$

which is recognized as valid in virtue of the computation rule

$$\frac{\begin{array}{l}\mathrm{el}((\Pi\, x : A)B \,|\, \gamma) : f \,|\, \gamma \Rightarrow g \;\varepsilon\; \mathrm{el}(\Pi(C, \mathrm{cl}(\hat{x}B, \gamma))) \\ \mathrm{el}(A \,|\, \gamma) : a \,|\, \gamma \Rightarrow c \;\varepsilon\; \mathrm{el}(C) \\ \mathrm{el}(\mathrm{app}[\mathrm{cl}(\hat{x}B, \gamma), c]) : \mathrm{app}[g, c] \Rightarrow d \;\varepsilon\; \mathrm{el}(D)\end{array}}{\mathrm{el}(B \circ (\mathrm{id}, x \leftarrow a) \,|\, \gamma) : \mathrm{app}(f, a) \,|\, \gamma \Rightarrow d \;\varepsilon\; \mathrm{el}(D)}, \tag{C5.31}$$

i.e., applications are evaluated eagerly. The reader is advised to check that, if c is the value of $a \,|\, \gamma$, then $B \circ (\mathrm{id}, x \leftarrow a) \,|\, \gamma$ and $\mathrm{app}[\mathrm{cl}(\hat{x}B, \gamma), c]$ evaluate to the same set D. Two elements of the form $\mathrm{app}(f, a)$ are equal if their parts are equal:

$$\frac{f = g : \mathrm{el}((\Pi\, x : A)B)\ (\Gamma) \quad a = b : \mathrm{el}(A)\ (\Gamma)}{\mathrm{app}(f, a) = \mathrm{app}(g, b) : \mathrm{el}(B \circ (\mathrm{id}, x \leftarrow b))\ (\Gamma)} \tag{J5.72}$$

Note that $B \circ (\mathrm{id}, x \leftarrow a) = B \circ (\mathrm{id}, x \leftarrow b)) : \mathrm{set}\ (\Gamma)$.

Justification. Let the premisses be given and let γ be an arbitrarily given canonical assignment for Γ. The computation of $\mathrm{app}(f, a) \,|\, \gamma$ involves set conversion:

$$\frac{\dfrac{\begin{array}{l}\mathrm{el}((\Pi\, x : A)B \,|\, \gamma) : f \,|\, \gamma \Rightarrow f_0 \;\varepsilon\; \mathrm{el}(\Pi(A_0, \mathrm{cl}(\hat{x}B, \gamma))) \\ \mathrm{el}(A \,|\, \gamma) : a \,|\, \gamma \Rightarrow a_0 \;\varepsilon\; \mathrm{el}(A_0) \\ \mathrm{el}(\mathrm{app}[\mathrm{cl}(\hat{x}B, \gamma), a_0]) : \mathrm{app}[f_0, a_0] \Rightarrow d \;\varepsilon\; \mathrm{el}(D)\end{array}}{\mathrm{el}(B \circ (\mathrm{id}, x \leftarrow a) \,|\, \gamma) : \mathrm{app}(f, a) \,|\, \gamma \Rightarrow d \;\varepsilon\; \mathrm{el}(D)} \quad \begin{array}{l} B \circ (\mathrm{id}, x \leftarrow a) = \\ \quad B \circ (\mathrm{id}, x \leftarrow b) : \mathrm{set} \\ B \circ (\mathrm{id}, x \leftarrow b) \Rightarrow E \;\varepsilon\; \mathrm{set}\end{array}}{\mathrm{el}(B \circ (\mathrm{id}, x \leftarrow b) \,|\, \gamma) : \mathrm{app}(f, a) \,|\, \gamma \Rightarrow d \;\varepsilon\; \mathrm{el}(E)},$$

and the computation of $\mathrm{app}(g, b) \,|\, \gamma$ is given by

$$\frac{\begin{array}{l}\mathrm{el}((\Pi\, x : A)B \,|\, \gamma) : g \,|\, \gamma \Rightarrow g_0 \;\varepsilon\; \mathrm{el}(\Pi(A_0, \mathrm{cl}(\hat{x}B, \gamma))) \\ \mathrm{el}(A \,|\, \gamma) : b \,|\, \gamma \Rightarrow b_0 \;\varepsilon\; \mathrm{el}(A_0) \\ \mathrm{el}(\mathrm{app}[\mathrm{cl}(\hat{x}B, \gamma), b_0]) : \mathrm{app}[g_0, b_0] \Rightarrow e \;\varepsilon\; \mathrm{el}(E)\end{array}}{\mathrm{el}(B \circ (\mathrm{id}, x \leftarrow b) \,|\, \gamma) : \mathrm{app}(g, b) \,|\, \gamma \Rightarrow e \;\varepsilon\; \mathrm{el}(E)}.$$

It remains to show that $d = e \;\varepsilon\; \mathrm{el}(E)$. This demonstration proceeds in three steps: first, $a_0 = b_0 \;\varepsilon\; \mathrm{el}(A_0)$,

$$\frac{\dfrac{a = b : \mathrm{el}(A)\ (\Gamma) \quad \gamma \;\varepsilon\; \mathrm{ass}(\Gamma)}{a \,|\, \gamma = b \,|\, \gamma : \mathrm{el}(A \,|\, \gamma)} \quad \begin{array}{l}\mathrm{el}(A \,|\, \gamma) : a \,|\, \gamma \Rightarrow a_0 \;\varepsilon\; \mathrm{el}(A_0) \\ \mathrm{el}(A \,|\, \gamma) : b \,|\, \gamma \Rightarrow b_0 \;\varepsilon\; \mathrm{el}(A_0)\end{array}}{a_0 = b_0 \;\varepsilon\; \mathrm{el}(A_0)};$$

next, $f_0 = g_0 \ \varepsilon \ \mathrm{el}(\Pi(A_0, \mathrm{cl}(\hat{x}B, \gamma)))$,

$$\dfrac{\dfrac{f = g : \mathrm{el}((\Pi\, x : A)B)\ (\Gamma) \quad \gamma \ \varepsilon \ \mathrm{ass}(\Gamma)}{f \,|\, \gamma = g \,|\, \gamma : \mathrm{el}((\Pi\, x : A)B \,|\, \gamma)} \quad \begin{array}{l}\mathrm{el}((\Pi\, x : A)B \,|\, \gamma) : f \,|\, \gamma \Rightarrow f_0 \ \varepsilon \ \mathrm{el}(G) \\ \mathrm{el}((\Pi\, x : A)B \,|\, \gamma) : g \,|\, \gamma \Rightarrow g_0 \ \varepsilon \ \mathrm{el}(G)\end{array}}{f_0 = g_0 \ \varepsilon \ \mathrm{el}(\Pi(A_0, \mathrm{cl}(\hat{x}B, \gamma)))},$$

where, for typographical reasons, G abbreviates $\Pi(A_0, \mathrm{cl}(\hat{x}B, \gamma))$; and, finally,

$$\dfrac{\dfrac{\begin{array}{l} f_0 = g_0 \ \varepsilon \ \mathrm{el}(\Pi(A_0, \mathrm{cl}(\hat{x}B, \gamma))) \\ a_0 = b_0 \ \varepsilon \ \mathrm{el}(A_0)\end{array}}{\mathrm{app}[f_0, a_0] = \mathrm{app}[g_0, b_0] : \mathrm{el}(\mathrm{app}[\mathrm{cl}(\hat{x}B, \gamma), b_0])} \quad \begin{array}{l}\mathrm{el}(F) : \mathrm{app}[f_0, a_0] \Rightarrow d \ \varepsilon \ \mathrm{el}(E) \\ \mathrm{el}(F) : \mathrm{app}[g_0, b_0] \Rightarrow e \ \varepsilon \ \mathrm{el}(E)\end{array}}{d = e \ \varepsilon \ \mathrm{el}(E)}$$

where, again for typographical reasons, F abbreviates $\mathrm{app}[\mathrm{cl}(\hat{x}B, \gamma), b_0]$. This completes the justification.

Since no variable binding operations are involved, substitutions can be moved in under application, i.e., the inference rule

$$\frac{f : \mathrm{el}((\Pi\, x : A)B)\ (\Gamma) \quad a : \mathrm{el}(A)\ (\Gamma) \quad \sigma : \Gamma \longleftarrow \Delta}{\mathrm{app}(f, a) \circ \sigma = \mathrm{app}(f \circ \sigma, a \circ \sigma) : \mathrm{el}(B \circ (\mathrm{id}, x \leftarrow a) \circ \sigma)\ (\Delta)} \tag{J5.73}$$

is valid. Note that, under the premisses of this inference rule,

$$\mathrm{app}(f \circ \sigma, a \circ \sigma) : \mathrm{el}(B \circ (\sigma \circ \mathrm{p}, x \leftarrow y) \circ (\mathrm{id}, y \leftarrow a \circ \sigma))\ (\Delta),$$

and that

$$\begin{array}{l}(\sigma \circ \mathrm{p}, x \leftarrow y) \circ (\mathrm{id}, y \leftarrow a \circ \sigma) = \\ (\sigma \circ \mathrm{p} \circ (\mathrm{id}, y \leftarrow a \circ \sigma), x \leftarrow y \circ (\mathrm{id}, y \leftarrow a \circ \sigma)) = \\ (\sigma \circ \mathrm{id}, x \leftarrow a \circ \sigma) = \\ (\sigma, x \leftarrow a \circ \sigma) = (\mathrm{id}, x \leftarrow a) \circ \sigma : (\Gamma, x : \mathrm{el}(A)) \longleftarrow \Delta,\end{array}$$

so the inference rule is well-formed.

Justification of (J5.73). Let the premisses be given and let δ be an arbitrarily given canonical assignment for Δ. Compare the computation traces

$$\dfrac{\sigma \,|\, \delta \Rightarrow \gamma \ \varepsilon \ \mathrm{ass}(\Gamma) \quad \dfrac{\begin{array}{l}\mathrm{el}((\Pi\, x : A)B \,|\, \gamma) : f \,|\, \gamma \Rightarrow g \ \varepsilon \ \mathrm{el}(\Pi(C, \mathrm{cl}(\hat{x}B, \gamma))) \\ \mathrm{el}(A \,|\, \gamma) : a \,|\, \gamma \Rightarrow c \ \varepsilon \ \mathrm{el}(C) \\ \mathrm{el}(\mathrm{app}[\mathrm{cl}(\hat{x}B, \gamma), c]) : \mathrm{app}[g, c] \Rightarrow d \ \varepsilon \ \mathrm{el}(D)\end{array}}{\mathrm{el}(B \circ (\mathrm{id}, x \leftarrow a) \,|\, \gamma) : \mathrm{app}(f, a) \,|\, \gamma \Rightarrow d \ \varepsilon \ \mathrm{el}(D)}}{\mathrm{el}(B \circ (\mathrm{id}, x \leftarrow a) \circ \sigma \,|\, \delta) : \mathrm{app}(f, a) \circ \sigma \,|\, \delta \Rightarrow d \ \varepsilon \ \mathrm{el}(D)}$$

and

$$\dfrac{\dfrac{\begin{array}{l}(\Pi\, x : A \circ \sigma)B \circ (\sigma \circ \mathrm{p}, x \leftarrow y) \,|\, \delta : \\ \quad f \circ \sigma \,|\, \delta \Rightarrow g \ \varepsilon \ \mathrm{el}(\Pi(C, \mathrm{cl}(\hat{y}B \circ (\sigma \circ \mathrm{p}, x \leftarrow y), \delta))) \\ \mathrm{el}(A \circ \sigma \,|\, \delta) : a \circ \sigma \,|\, \delta \Rightarrow c \ \varepsilon \ \mathrm{el}(C) \\ \mathrm{el}(\mathrm{app}[\mathrm{cl}(\hat{y}B \circ (\sigma \circ \mathrm{p}, x \leftarrow y), \delta), c]) : \mathrm{app}[g, c] \Rightarrow d \ \varepsilon \ \mathrm{el}(D)\end{array}}{\mathrm{el}(B \circ (\sigma \circ \mathrm{p}, x \leftarrow y) \circ (\mathrm{id}, y \leftarrow a \circ \sigma) \,|\, \delta) : \mathrm{app}(f \circ \sigma, a \circ \sigma) \,|\, \delta \Rightarrow d \ \varepsilon \ \mathrm{el}(D)}}{\mathrm{el}(B \circ (\mathrm{id}, x \leftarrow a) \circ \sigma \,|\, \delta) : \mathrm{app}(f \circ \sigma, a \circ \sigma) \,|\, \delta \Rightarrow d \ \varepsilon \ \mathrm{el}(D)}.$$

In the latter computation trace, only the last step is displayed for typographical reasons. Recall that $\mathrm{app}[\mathrm{cl}(\hat{y}B \circ (\sigma \circ \mathrm{p}, x \leftarrow y), \delta), c]$ and $\mathrm{app}[\mathrm{cl}(\hat{x}B, \gamma), c]$ have the same value, if $\sigma \,|\, \delta \Rightarrow \gamma \ \varepsilon \ \mathrm{ass}(\Gamma)$. This completes the justification.

If we take σ to be id in (J5.71), we get the inference rule

$$\frac{b : \mathrm{el}(B)\ (\Gamma, x : \mathrm{el}(A))}{(\lambda\, x)b = (\lambda\, y)(b \circ (\mathrm{p}, x \leftarrow y)) : \mathrm{el}((\Pi\, x : A)B)\ (\Gamma)} \tag{α}$$

which is the rule of α-conversion in the present setting.

Justification of (β). Let γ be an arbitrary canonical assignment for the context Γ. As usual, the equality between the two terms of the conclusion is justified by comparing their computation traces, i.e.,

$$\frac{\mathrm{el}((\Pi\, x : A)B \,|\, \gamma) : (\lambda\, x)b \,|\, \gamma \Rightarrow \mathrm{cl}(\hat{x}b, \gamma)\ \varepsilon\ \mathrm{el}(\Pi(C, \mathrm{cl}(\hat{x}b, \gamma))) \quad \mathrm{el}(A \,|\, \gamma) : a \,|\, \gamma \Rightarrow c\ \varepsilon\ \mathrm{el}(C) \quad \dfrac{\mathrm{el}(B \,|(\gamma, x \leftarrow c)) : b \,|(\gamma, x \leftarrow c) \Rightarrow d\ \varepsilon\ \mathrm{el}(D)}{\mathrm{el}(\mathrm{app}[\mathrm{cl}(\hat{x}B, \gamma), c]) : \mathrm{app}[\mathrm{cl}(\hat{x}b, \gamma), c] \Rightarrow d\ \varepsilon\ \mathrm{el}(D)}}{\mathrm{el}(B \circ (\mathrm{id}, x \leftarrow a) \,|\, \gamma) : \mathrm{app}((\lambda\, x)b, a) \,|\, \gamma \Rightarrow d\ \varepsilon\ \mathrm{el}(D)}$$

and

$$\frac{\dfrac{\mathrm{id} \,|\, \gamma \Rightarrow \gamma\ \varepsilon\ \mathrm{ass}(\Gamma) \quad \mathrm{el}(A \,|\, \gamma) : a \,|\, \gamma \Rightarrow c\ \varepsilon\ \mathrm{el}(C)}{(\mathrm{id}, x \leftarrow a) \,|\, \gamma \Rightarrow (\gamma, x \leftarrow c)\ \varepsilon\ \mathrm{ass}(\Gamma, x : \mathrm{el}(A))} \quad \mathrm{el}(B \,|(\gamma, x \leftarrow c)) : b \,|(\gamma, x \leftarrow c) \Rightarrow d\ \varepsilon\ \mathrm{el}(D)}{\mathrm{el}(B \circ (\mathrm{id}, x \leftarrow a) \,|\, \gamma) : b \circ (\mathrm{id}, x \leftarrow a) \,|\, \gamma \Rightarrow d\ \varepsilon\ \mathrm{el}(D)}\ .$$

They show that $\mathrm{app}((\lambda\, x)b, a) \,|\, \gamma$ and $b \circ (\mathrm{id}, x \leftarrow a) \,|\, \gamma$ are equal elements of the set $B \circ (\mathrm{id}, x \leftarrow a) \,|\, \gamma$, as required. This completes the justification.

Justification of (η). Let γ be an arbitrary canonical assignment for the context Γ. It has to be shown that $f \,|\, \gamma$ and $(\lambda\, x)\,\mathrm{app}(f \circ \mathrm{p}, x) \,|\, \gamma$ are equal elements of the set $(\Pi\, x : A)B \,|\, \gamma$. Let $A \,|\, \gamma$ have the canonical set C as value and let $f \,|\, \gamma$ have the canonical element g of the canonical set $\Pi(C, \mathrm{cl}(\hat{x}B, \gamma))$ as value; note that the value of $(\lambda\, x)\,\mathrm{app}(f \circ \mathrm{p}, x) \,|\, \gamma$ is $\mathrm{cl}(\hat{x}\,\mathrm{app}(f \circ \mathrm{p}, x), \gamma)$. It has to be shown that g and $\mathrm{cl}(\hat{x}\,\mathrm{app}(f \circ \mathrm{p}, x), \gamma)$ are equal canonical elements of the set $\Pi(C, \mathrm{cl}(\hat{x}B, \gamma))$. Thus, let c be an arbitrary canonical element of C and assume that the value of $\mathrm{app}[g, c]$ is the canonical element d of the canonical set D which is the value of $B \,|(\gamma, x \leftarrow c)$, or, which amounts to the same, of $\mathrm{app}[\mathrm{cl}(\hat{x}B, \gamma), c]$. It remains to show that, under these assumptions, the value of $\mathrm{app}[\mathrm{cl}(\hat{x}\,\mathrm{app}(f \circ \mathrm{p}, x), \gamma), c]$ is also d. The full computation trace is given by

$$\frac{\dfrac{\dfrac{\mathrm{p} \,|(\gamma, x \leftarrow c) \Rightarrow \gamma\ \varepsilon\ \mathrm{ass}(\Gamma) \quad \mathrm{el}((\Pi\, x : A)B \,|\, \gamma) : f \,|\, \gamma \Rightarrow g\ \varepsilon\ \mathrm{el}(\Pi(C, \mathrm{cl}(\hat{x}B, \gamma)))}{\mathrm{el}(((\Pi\, x : A)B) \circ \mathrm{p} \,|(\gamma, x \leftarrow c)) : (f \circ \mathrm{p}) \,|(\gamma, x \leftarrow c) \Rightarrow g\ \varepsilon\ \mathrm{el}(\Pi(C, \mathrm{cl}(\hat{x}B, \gamma)))} \quad \begin{matrix} \mathrm{el}(A \circ \mathrm{p} \,|(\gamma, x \leftarrow c)) : x \,|(\gamma, x \leftarrow c) \Rightarrow c\ \varepsilon\ \mathrm{el}(C) \\ \mathrm{el}(\mathrm{app}[\mathrm{cl}(\hat{x}B, \gamma), c]) : \mathrm{app}[g, c] \Rightarrow d\ \varepsilon\ \mathrm{el}(D) \end{matrix}}{\mathrm{el}(B \,|(\gamma, x \leftarrow c)) : \mathrm{app}(f \circ \mathrm{p}, x) \,|(\gamma, x \leftarrow c) \Rightarrow d\ \varepsilon\ \mathrm{el}(D)}}{\mathrm{el}(\mathrm{app}[\mathrm{cl}(\hat{x}B, \gamma), c]) : \mathrm{app}[\mathrm{cl}(\hat{x}\,\mathrm{app}(f \circ \mathrm{p}, x), \gamma), c] \Rightarrow d\ \varepsilon\ \mathrm{el}(D)}\ .$$

This completes the justification.

A very important inference rule, connecting application and λ abstraction, is the rule of β-conversion:

$$\frac{b : \mathrm{el}(B)\ (\Gamma, x : \mathrm{el}(A)) \quad a : \mathrm{el}(A)\ (\Gamma)}{\mathrm{app}((\lambda\, x)b, a) = b \circ (\mathrm{id}, x \leftarrow a) : \mathrm{el}(B \circ (\mathrm{id}, x \leftarrow a))\ (\Gamma)}\ . \tag{β}$$

Its justification is given on a separate page (p. 141).

Since the set $(\Pi\, x : A)B$ has been defined by its elimination rule, the rule of η-conversion can also be justified—something which is not possible if the set $(\Pi\, x : A)B$ is given an inductive definition; note that the higher type structure is required to give the set $(\Pi\, x : A)B$ an inductive definition.

$$\frac{f : \mathrm{el}((\Pi\, x : A)B)\ (\Gamma)}{f = (\lambda\, x)\, \mathrm{app}(f \circ \mathrm{p}, x) : \mathrm{el}((\Pi\, x : A)B)\ (\Gamma)} \tag{η}$$

Its justification is given on a separate page (p. 141).

This completes the treatment of the set $(\Pi\, x : A)B$.

The set $A \to B$ of functions from A to B, introduced in Chapter IV, Section 6, can be defined in terms of Π, as the special case when B does not depend on x:

$$\frac{A : \mathrm{set}\ (\Gamma) \quad B : \mathrm{set}\ (\Gamma)}{A \to B \stackrel{\mathrm{def}}{=} (\Pi\, x : A)B \circ \mathrm{p} : \mathrm{set}\ (\Gamma)}\ . \tag{Df}$$

As explained in Chapter IV, Section 5, such a definition is to be understood as an abbreviation for the three inference rules

$$\frac{A : \mathrm{set}\ (\Gamma) \quad B : \mathrm{set}\ (\Gamma)}{A \to B : \mathrm{set}\ (\Gamma)}\ , \tag{R5.32}$$

$$\frac{A \Rightarrow C\ \varepsilon\ \mathrm{set}}{A \to B \,|\, \gamma \Rightarrow \Pi(C, \mathrm{cl}(\hat{x}B \circ \mathrm{p}))\ \varepsilon\ \mathrm{set}}\ , \tag{C5.32}$$

and

$$\frac{A : \mathrm{set}\ (\Gamma) \quad B : \mathrm{set}\ (\Gamma)}{A \to B = (\Pi\, x : A)B \circ \mathrm{p} : \mathrm{set}\ (\Gamma)}\ . \tag{J5.74}$$

In virtue of this nominal definition, the same λ and app work for both $A \to B$ and $(\Pi\, x : A)B \circ \mathrm{p}$.

§ 6. The disjoint union of a family of sets

The set $(\Sigma\, x : A)B$, introduced in Chapter IV, Section 7, also has to be generalized to the hypothetical case. The formation rule is given by

$$\frac{A : \mathrm{set}\ (\Gamma) \quad B : \mathrm{set}\ (\Gamma, x : \mathrm{el}(A))}{(\Sigma\, x : A)B : \mathrm{set}\ (\Gamma)}\ . \tag{R5.33}$$

This inference rule is recognized as valid in virtue of the computation rule

$$\frac{A \,|\, \gamma \Rightarrow C\ \varepsilon\ \mathrm{set}}{(\Sigma\, x : A)B \,|\, \gamma \Rightarrow \Sigma(C, \mathrm{cl}(\hat{x}B, \gamma))\ \varepsilon\ \mathrm{set}}\ . \tag{C5.33}$$

As usual, it has to be checked that two sets of this form are equal if their parts are equal:

$$\frac{A = C : \mathrm{set}\ (\Gamma) \quad B = D : \mathrm{set}\ (\Gamma, x : \mathrm{el}(A))}{(\Sigma\, x : A)B = (\Sigma\, x : C)D : \mathrm{set}\ (\Gamma)}\ . \tag{J5.75}$$

The justification is analogous to that of (J5.67), using (J4.14) instead of (J4.15). In the same way, the justification of the inference rule

$$\frac{A : \mathrm{set}\ (\Gamma) \quad B : \mathrm{set}\ (\Gamma, x : \mathrm{el}(A)) \quad \sigma : \Gamma \longleftarrow \Delta}{((\Sigma\, x : A)B) \circ \sigma = (\Sigma\, y : A \circ \sigma)(B \circ (\sigma \circ \mathrm{p}, x \leftarrow y)) : \mathrm{set}\ (\Delta)} \qquad \text{(J5.76)}$$

is analogous to that of (J5.68)

The introduction rule for hypothetical elements of the Σ set is given by

$$\frac{\begin{array}{c}(B : \mathrm{set}\ (\Gamma, x : \mathrm{el}(A)))\\ a : \mathrm{el}(A)\ (\Gamma) \quad b : \mathrm{el}(B \circ (\mathrm{id}, x \leftarrow a))\ (\Gamma)\end{array}}{(a\,, b) : \mathrm{el}((\Sigma\, x : A)B)\ (\Gamma)}\ . \qquad \text{(R5.34)}$$

This inference rule is recognized as valid in virtue of the computation rule

$$\frac{\begin{array}{l}\mathrm{el}(A \,|\, \gamma) : a \,|\, \gamma \Rightarrow c \ \varepsilon\ \mathrm{el}(C)\\ \mathrm{el}(B \circ (\mathrm{id}, x \leftarrow a) \,|\, \gamma) : b \,|\, \gamma \Rightarrow d \ \varepsilon\ \mathrm{el}(D)\end{array}}{\mathrm{el}((\Sigma x : A)B \,|\, \gamma) : (a\,, b) \,|\, \gamma \Rightarrow (c\,, d) \ \varepsilon\ \mathrm{el}(\Sigma(C, \mathrm{cl}(\hat{x}B, \gamma)))}\ . \qquad \text{(C5.34)}$$

Note that $B \circ (\mathrm{id}, x \leftarrow a) \,|\, \gamma = \mathrm{app}[\mathrm{cl}(\hat{x}B, \gamma), c] : \mathrm{set}$, under the first premiss of this computation rule.

$$\frac{\begin{array}{c}(B : \mathrm{set}\ (\Gamma, x : \mathrm{el}(A)))\\ a = c : \mathrm{el}(A)\ (\Gamma) \quad b = d : \mathrm{el}(B \circ (\mathrm{id}, x \leftarrow a))\ (\Gamma)\end{array}}{(a\,, b) = (c, d) : \mathrm{el}((\Sigma\, x : A)B)\ (\Gamma)}\ . \qquad \text{(J5.77)}$$

Justification. Let the premisses be given and let γ be an arbitrarily given canonical assignment for Γ. Furthermore, let the values of $a \,|\, \gamma$, $b \,|\, \gamma$, $c \,|\, \gamma$, and $d \,|\, \gamma$ be a_0, b_0, c_0 and d_0, respectively of the relevant sets. By (D5.14) and (D4.2), $a_0 = b_0 \ \varepsilon\ \mathrm{el}(A_0)$, where A_0 is the value of $A \,|\, \gamma$; by the same inference rules, $c_0 = d_0 \ \varepsilon\ \mathrm{el}(B_0)$, where B_0 is the value of $B \circ (\mathrm{id}, x \leftarrow a) \,|\, \gamma$. Now B_0 is also the value of $\mathrm{app}[\mathrm{cl}(\hat{x}B, \gamma), a_0]$ and the result follows from (D4.10).

$$\frac{\begin{array}{ll}a : \mathrm{el}(A)\ (\Gamma) & b : \mathrm{el}(B \circ (\mathrm{id}, x \leftarrow a))\ (\Gamma)\\ B : \mathrm{set}\ (\Gamma, x : \mathrm{el}(A)) & \sigma : \Gamma \longleftarrow \Delta\end{array}}{(a\,, b) \circ \sigma = (a \circ \sigma, b \circ \sigma) : \mathrm{el}(((\Sigma\, x : A)B) \circ \sigma)\ (\Delta)}\ . \qquad \text{(J5.78)}$$

From the premisses, it follows that $b \circ \sigma : \mathrm{el}(B \circ (\mathrm{id}, x \leftarrow a) \circ \sigma)\ (\Delta)$; on the other hand, $b \circ \sigma$ is expected to be an element of the set $B \circ (\sigma \circ \mathrm{p}, x \leftarrow y) \circ (\mathrm{id}, y \leftarrow a \circ \sigma)$ for $(a \circ \sigma, b \circ \sigma)$ to be an element of the set $(\Sigma\, y : A \circ \sigma)B \circ (\sigma \circ \mathrm{p}, x \leftarrow y)$ which is equal to the set $((\Sigma\, x : A)B) \circ \sigma$, all this in the context Δ; but

$$(\sigma \circ \mathrm{p}, x \leftarrow y) \circ (\mathrm{id}, y \leftarrow a \circ \sigma) = (\mathrm{id}, x \leftarrow a) \circ \sigma : (\Gamma, x : \mathrm{el}(A)) \longleftarrow \Delta,$$

as shown on p. 140, so $b \circ \sigma$ is indeed an element of the correct set.

Justification of (J5.78). Let the (four) premisses be given and let δ be an arbitrarily given canonical assignment for Δ. Compare the computation trace

$$\frac{\sigma \,|\, \delta \Rightarrow \gamma \ \varepsilon\ \mathrm{ass}(\Gamma) \quad \dfrac{\begin{array}{l}\mathrm{el}(A \,|\, \gamma) : a \,|\, \gamma \Rightarrow a_0 \ \varepsilon\ \mathrm{el}(C)\\ \mathrm{el}(B \circ (\mathrm{id}, x \leftarrow a) \,|\, \gamma) : b \,|\, \gamma \Rightarrow b_0 \ \varepsilon\ \mathrm{el}(D)\end{array}}{\mathrm{el}((\Sigma\, x : A)B \,|\, \gamma) : (a\,, b) \,|\, \gamma \Rightarrow (a_0\,, b_0) \ \varepsilon\ \mathrm{el}(\Sigma(C, \mathrm{cl}(\hat{x}B, \gamma)))}}{\mathrm{el}((\Sigma\, x : A)B \circ \sigma \,|\, \delta) : (a\,, b) \circ \sigma \,|\, \delta \Rightarrow (a_0\,, b_0) \ \varepsilon\ \mathrm{el}(\Sigma(C, \mathrm{cl}(\hat{x}B, \gamma)))}$$

to the computation trace

$$\dfrac{\dfrac{\begin{matrix}\sigma \mid \delta \Rightarrow \gamma \;\varepsilon\; \mathrm{ass}(\Gamma) \\ \mathrm{el}(A \mid \gamma) : a \mid \gamma \Rightarrow a_0 \;\varepsilon\; \mathrm{el}(C)\end{matrix}}{\mathrm{el}(A \circ \sigma \mid \delta) : a \circ \sigma \mid \delta \Rightarrow a_0 \;\varepsilon\; \mathrm{el}(C)} \qquad \dfrac{\dfrac{\begin{matrix}\sigma \mid \delta \Rightarrow \gamma \;\varepsilon\; \mathrm{ass}(\Gamma) \\ \mathrm{el}(B \circ (\mathrm{id}, x \leftarrow a) \mid \gamma) : b \mid \gamma \Rightarrow b_0 \;\varepsilon\; \mathrm{el}(D)\end{matrix}}{\mathrm{el}(B \circ (\mathrm{id}, x \leftarrow a) \circ \sigma \mid \delta) : b \circ \sigma \mid \delta \Rightarrow b_0 \;\varepsilon\; \mathrm{el}(D)}}{\begin{matrix}\mathrm{el}(B \circ (\sigma \circ \mathrm{p}, x \leftarrow y) \circ (\mathrm{id}, y \leftarrow a \circ \sigma) \mid \delta) \\ : b \circ \sigma \mid \delta \Rightarrow b_0 \;\varepsilon\; \mathrm{el}(D)\end{matrix}}}{\dfrac{\begin{matrix}\mathrm{el}((\Sigma x : A \circ \sigma) B \circ (\sigma \circ \mathrm{p}, x \leftarrow y) \mid \delta) : \\ (a \circ \sigma, b \circ \sigma) \mid \delta \Rightarrow (a_0, b_0) \;\varepsilon\; \mathrm{el}(\Sigma(C, \mathrm{cl}(\hat{x} B \circ (\sigma \circ \mathrm{p}, x \leftarrow y), \delta)))\end{matrix}}{\mathrm{el}((\Sigma x : A) B \circ \sigma \mid \delta) : (a \circ \sigma, b \circ \sigma) \mid \delta \Rightarrow (a_0, b_0) \;\varepsilon\; \mathrm{el}(\Sigma(C, \mathrm{cl}(\hat{x} B, \gamma)))}} \, .$$

In the second computation trace, there are two applications of (C4.1), i.e., set conversion, without mention of the minor premiss, but this is clear from the context.

§ 7. Elimination rules

Several ways of defining functions have been proposed in this book: for example, addition was informally defined on p. 32; on p. 84, the functional form dbl was defined by an analytic expression, and on p. 86 the same function was redefined by computation rules.

The most fundamental way of defining a function is by its computation rules. Some basic functions are needed before new functions can be defined in terms of these basic functions by analytic expressions, and the basic functions have to be defined by their computation rules: so, at least in this sense, the definition of functions by their computation rules is prior to the definition of functions by analytic expressions.

Consider now the inference rule

$$\frac{a : \mathrm{el}(\mathrm{N})\ (\Gamma)}{\mathrm{dbl}(a) : \mathrm{el}(\mathrm{N})\ (\Gamma)}$$

for the double function. As before, this inference rule is recognized as valid in virtue of the computation rule

$$\frac{\mathrm{el}(\mathrm{N} \mid \gamma) : a \mid \gamma \Rightarrow b \;\varepsilon\; \mathrm{el}(\mathrm{N}) \quad \mathrm{el}(\mathrm{N}) : \mathrm{dbl}[b] \Rightarrow c \;\varepsilon\; \mathrm{el}(\mathrm{N})}{\mathrm{el}(\mathrm{N} \mid \gamma) : \mathrm{dbl}(a) \mid \gamma \Rightarrow c \;\varepsilon\; \mathrm{el}(\mathrm{N})} \, ,$$

which, in its turn, relies on the inference rule

$$\frac{a \;\varepsilon\; \mathrm{el}(\mathrm{N})}{\mathrm{dbl}[a] : \mathrm{el}(\mathrm{N})}$$

recognized as valid in virtue of the computation rules

$$\mathrm{el}(\mathrm{N}) : \mathrm{dbl}[0] \Rightarrow 0 \;\varepsilon\; \mathrm{el}(\mathrm{N})$$

and

$$\frac{\mathrm{el}(\mathrm{N}) : \mathrm{dbl}[a] \Rightarrow b \;\varepsilon\; \mathrm{el}(\mathrm{N})}{\mathrm{el}(\mathrm{N}) : \mathrm{dbl}[\mathrm{s}(a)] \Rightarrow \mathrm{s}(\mathrm{s}(b)) \;\varepsilon\; \mathrm{el}(\mathrm{N})} \, .$$

This way of defining functions soon becomes repetitive and impeding, but I think it is important to point out that it *is feasible*.

The more convenient and economical approach is to introduce schemata for defining functions: typically one schema for each form of set will be sufficient. Since nominal definitions of functional forms were clarified above, it suffices to introduce schemata for defining functions of variables. Such schemata resemble the *special forms* of conventional programming languages, e.g., the "if... then... else..." form. Such forms are called *special* in conventional programming languages since they do not obey the usual operational semantic rules of these languages (e.g., eager evaluation of arguments from left to right); they have a *special* semantic also in intuitionistic type theory, but this shows itself only in the computation rules, not in the inference rules governing them.

Most interesting aspects of elimination rules are present already for the natural numbers. Just as in Section 4 of this chapter, I will consider only the natural numbers in detail. For the other sets and set forming operations, the inference rules are only given—without justifications.

$$\frac{\begin{array}{l} n : \mathrm{el}(\mathrm{N})\ (\Gamma) \\ C : \mathrm{set}\ (\Gamma, x : \mathrm{el}(\mathrm{N})) \\ a : \mathrm{el}(C \circ (\mathrm{id}, x \leftarrow 0))\ (\Gamma) \\ b : \mathrm{el}(C \circ (\mathrm{p}, x \leftarrow \mathrm{s}(y)) \circ \mathrm{p})\ (\Gamma, y : \mathrm{el}(\mathrm{N}), z : \mathrm{el}(C \circ (\mathrm{p}, x \leftarrow y))) \end{array}}{\mathrm{R}_{\hat{x}C}(n, a, \hat{y}\hat{z}b) : \mathrm{el}(C \circ (\mathrm{id}, x \leftarrow n))\ (\Gamma)}\ .^{14} \qquad \text{(R5.35)}$$

and that y and z become bound in b. The reader is advised to check that this inference rule is well-formed; here the mediate inference rules (M5.14) and (M5.16) are of benefit. To compute $\mathrm{R}_{\hat{x}C}(n, a, \hat{y}\hat{z}b) \,|\, \gamma$, for an arbitrary given assignment γ for Γ, first $n \,|\, \gamma$ is computed, and then the computation continues with a or b depending on whether the value of $n \,|\, \gamma$ is zero or not.

To formulate the above in terms of computation rules, I will use the intermediate form of expression $\mathrm{R}_{\hat{x}C}[m](a, \hat{y}\hat{z}b) \,|\, \gamma$, in which m is the value of $n \,|\, \gamma$, i.e.,

$$\frac{\begin{array}{l} \mathrm{el}(\mathrm{N} \,|\, \gamma) : n \,|\, \gamma \Rightarrow m \ \varepsilon\ \mathrm{el}(\mathrm{N}) \\ \mathrm{el}(C \,|(\gamma, x \leftarrow m)) : \mathrm{R}_{\hat{x}C}[m](a, \hat{y}\hat{z}b) \,|\, \gamma \Rightarrow d \ \varepsilon\ \mathrm{el}(D) \end{array}}{\mathrm{el}(C \circ (\mathrm{id}, x \leftarrow n) \,|\, \gamma) : \mathrm{R}_{\hat{x}C}(n, a, \hat{y}\hat{z}b) \,|\, \gamma \Rightarrow d \ \varepsilon\ \mathrm{el}(D)}\ . \qquad \text{(C5.35)}$$

It remains to show why $\mathrm{R}_{\hat{x}C}[m](a, \hat{y}\hat{z}b) \,|\, \gamma$ is a noncanonical element of the set $C \,|(\gamma, x \leftarrow m)$, i.e., to give the computation rules in virtue of which the inference rule

$$\frac{\begin{array}{l} m \ \varepsilon\ \mathrm{el}(\mathrm{N}) \qquad\qquad \gamma \ \varepsilon\ \mathrm{ass}(\Gamma) \\ C : \mathrm{set}\ (\Gamma, x : \mathrm{el}(\mathrm{N})) \\ a : \mathrm{el}(C \circ (\mathrm{id}, x \leftarrow 0))\ (\Gamma) \\ b : \mathrm{el}(C \circ (\mathrm{p}, x \leftarrow \mathrm{s}(y)) \circ \mathrm{p})\ (\Gamma, y : \mathrm{el}(\mathrm{N}), z : \mathrm{el}(C \circ (\mathrm{p}, x \leftarrow y))) \end{array}}{\mathrm{R}_{\hat{x}C}[m](a, \hat{y}\hat{z}b) \,|\, \gamma : \mathrm{el}(C \,|(\gamma, x \leftarrow m))} \qquad \text{(R5.36)}$$

is recognized as valid. As said above, this inference rule has two computation rules, one when m is zero, i.e.,

$$\frac{\mathrm{el}(C \circ (\mathrm{id}, x \leftarrow 0) \,|\, \gamma) : a \,|\, \gamma \Rightarrow d \ \varepsilon\ \mathrm{el}(D)}{\mathrm{el}(C \,|(\gamma, x \leftarrow 0)) : \mathrm{R}_{\hat{x}C}[0](a, \hat{y}\hat{z}b) \,|\, \gamma \Rightarrow d \ \varepsilon\ \mathrm{el}(D)}\ , \qquad \text{(C5.36)}$$

[14]The letter R is an abbreviation of *recursion*, since uses of this constant can be interpreted as definitions by recursion. This notation, as well as the notations D and E for the eliminatory constants of disjoint union and Cartesian product, are due to Martin-Löf, *Intuitionistic Type Theory*. Their definitions were also first given in the same book.

and one when m is non-zero, i.e.,

$$\frac{\begin{array}{l}\mathrm{el}(C\,|(\gamma,x\leftarrow p)):\mathrm{R}_{\hat{x}C}[p](a,\hat{y}\hat{z}b)\,|\,\gamma\Rightarrow e\ \varepsilon\ \mathrm{el}(E)\\ \mathrm{el}(C\circ(\mathrm{p},x\leftarrow \mathrm{s}(y))\circ\mathrm{p}\,|(\gamma,y\leftarrow p,z\leftarrow e)):\\ \qquad b\,|(\gamma,y\leftarrow p,z\leftarrow e)\Rightarrow d\ \varepsilon\ \mathrm{el}(D)\end{array}}{\mathrm{el}(C\,|(\gamma,x\leftarrow \mathrm{s}(p))):\mathrm{R}_{\hat{x}C}[\mathrm{s}(p)](a,\hat{y}\hat{z}b)\,|\,\gamma\Rightarrow d\ \varepsilon\ \mathrm{el}(D)}\ . \qquad \text{(C5.37)}$$

Note that

$$C\circ(\mathrm{id},x\leftarrow 0)\,|\,\gamma=C\,|(\gamma,x\leftarrow 0):\mathrm{set}$$

and that

$$C\circ(\mathrm{p},x\leftarrow \mathrm{s}(y))\circ\mathrm{p}\,|(\gamma,y\leftarrow p,z\leftarrow e)=C\,|(\gamma,x\leftarrow \mathrm{s}(p)):\mathrm{set},$$

so these inference rules are well-formed.

It is also worth noting that it is inference rules like (C5.37), and similar inference rules for other inductive sets, like well-orderings, that contribute to the strength of intuitionistic reasoning.[15]

Now, for each of the two forms of expression, it has to be checked that two terms of this form are equal if their parts are equal, and that a substitution can be moved in under the recursion operator. In the justifications of the following inference rules, set conversion will be implicit.

$$\frac{\begin{array}{l}n=m\ \varepsilon\ \mathrm{el}(\mathrm{N})\qquad \gamma=\delta\ \varepsilon\ \mathrm{ass}(\Gamma)\\ C=D:\mathrm{set}\ (\Gamma,x:\mathrm{el}(\mathrm{N}))\\ a=b:\mathrm{el}(D\circ(\mathrm{id},x\leftarrow 0))\ (\Gamma)\\ c=d:\mathrm{el}(D\circ(\mathrm{p},x\leftarrow \mathrm{s}(y))\circ\mathrm{p})\ (\Gamma,y:\mathrm{el}(\mathrm{N}),z:\mathrm{el}(D\circ(\mathrm{p},x\leftarrow y)))\end{array}}{\mathrm{R}_{\hat{x}C}[n](a,\hat{y}\hat{z}c)\,|\,\gamma=\mathrm{R}_{\hat{x}D}[m](b,\hat{y}\hat{z}d)\,|\,\delta:\mathrm{el}(D\,|(\delta,x\leftarrow m))}\ . \qquad \text{(J5.79)}$$

Justification. Let the premisses be given. Recall the definition of the canonical set N on p. 71 and consider two cases: either both m and n are zero, or m is $\mathrm{s}(p)$ and n is $\mathrm{s}(q)$ for some equal canonical numbers p and q. In the former case, the value of the left-hand side is the value of $a\,|\,\gamma$ and that of the right-hand side is that of $b\,|\,\delta$. By (M5.6), $a\,|\,\gamma$ and $b\,|\,\delta$ are equal elements of the relevant set, and, by (D4.2), their values are equal.

In the second case, it can be assumed to be already established that

$$\mathrm{R}_{\hat{x}C}[p](a,\hat{y}\hat{z}c)\,|\,\gamma=\mathrm{R}_{\hat{x}D}[q](b,\hat{y}\hat{z}d)\,|\,\delta:\mathrm{el}(D\,|(\delta,x\leftarrow q)),$$

so that the values of the two sides of this equality, say e and f, are equal elements of the set that is the value of $D\,|(\delta,x\leftarrow q)$. Now, because c and d are equal, γ and δ are equal, p and q are equal, and e and f are equal,

$$c\,|(\gamma,y\leftarrow p,z\leftarrow e)\qquad\text{and}\qquad d\,|(\delta,y\leftarrow q,z\leftarrow f)$$

are equal elements of the relevant set, by (D5.11), and (M5.6); it now follows from (C5.37) that

$$\mathrm{R}_{\hat{x}C}[\mathrm{s}(p)](a,\hat{y}\hat{z}c)\,|\,\gamma=\mathrm{R}_{\hat{x}D}[\mathrm{s}(q)](b,\hat{y}\hat{z}d)\,|\,\delta:\mathrm{el}(D\,|(\delta,x\leftarrow \mathrm{s}(q))),$$

as required. This completes the justification.

---

[15]The kind of reasoning adopted by Bellantoni and Cook ('A new recursion-theoretic characterization of the polytime functions') in their characterization of polynomial time computation suggests that, if b does not get access to the *canonical* value e in (C5.37), but only to its noncanonical counterpart, the resulting system becomes much weaker.

A similar inference rule can now be justified in the case when the first argument of $\mathrm{R}_{\hat{x}C}$ is not canonical:

$$\frac{\begin{array}{l} n = m : \mathrm{el}(\mathrm{N})\ (\Gamma) \\ C = D : \mathrm{set}\ (\Gamma, x : \mathrm{el}(\mathrm{N})) \\ a = b : \mathrm{el}(D \circ (\mathrm{id}, x \leftarrow 0))\ (\Gamma) \\ c = d : \mathrm{el}(D \circ (\mathrm{p}, x \leftarrow \mathrm{s}(y)) \circ \mathrm{p})\ (\Gamma, y : \mathrm{el}(\mathrm{N}), z : \mathrm{el}(D \circ (\mathrm{p}, x \leftarrow y))) \end{array}}{\mathrm{R}_{\hat{x}C}(n, a, \hat{y}\hat{z}c) = \mathrm{R}_{\hat{x}D}(m, b, \hat{y}\hat{z}d) : \mathrm{el}(D \circ (\mathrm{id}, x \leftarrow m))\ (\Gamma)} . \tag{J5.80}$$

The justification is immediate from (C5.35) and (J5.79).

The next inference rule shows how a substitution σ is moved in under the recursion operator.

$$\frac{\begin{array}{l} n : \mathrm{el}(\mathrm{N})\ (\Gamma) \\ C : \mathrm{set}\ (\Gamma, x : \mathrm{el}(\mathrm{N})) \\ a : \mathrm{el}(C \circ (\mathrm{id}, x \leftarrow 0))\ (\Gamma) \\ b : \mathrm{el}(C \circ (\mathrm{p}, x \leftarrow \mathrm{s}(y)) \circ \mathrm{p})\ (\Gamma, y : \mathrm{el}(\mathrm{N}), z : \mathrm{el}(C \circ (\mathrm{p}, x \leftarrow y))) \\ \sigma : \Gamma \longleftarrow \Delta \end{array}}{\begin{array}{l} \mathrm{R}_{\hat{x}C}(n, a, \hat{y}\hat{z}b) \circ \sigma = \\ \quad \mathrm{R}_{\hat{u}C \circ (\sigma \circ \mathrm{p}, x \leftarrow u)}(n \circ \sigma, a \circ \sigma, \hat{v}\hat{w}b \circ (\sigma \circ \mathrm{p} \circ \mathrm{p}, y \leftarrow v \circ \mathrm{p}, z \leftarrow w)) : \\ \quad \mathrm{el}(C \circ (\mathrm{id}, x \leftarrow n) \circ \sigma)\ (\Delta) \end{array}} . \tag{J5.81}$$

Justification. Let the premisses be given and let δ be an arbitrarily given canonical assignment for the context Δ. Let $\gamma \;\varepsilon\; \mathrm{ass}(\Gamma)$ be the value of $\sigma \mid \delta$ and let $m \;\varepsilon\; \mathrm{el}(\mathrm{N})$ be the value of $n \mid \gamma$. By set conversion, the value of the left-hand side is the same as the value of

$$\mathrm{R}_{\hat{x}C}[m](a, \hat{y}\hat{z}b) \mid \gamma,$$

and the value of the right-hand side is the same as the value of

$$\mathrm{R}_{\hat{u}C \circ (\sigma \circ \mathrm{p}, x \leftarrow u)}[m](a \circ \sigma, \hat{v}\hat{w}b \circ (\sigma \circ \mathrm{p} \circ \mathrm{p}, y \leftarrow v \circ \mathrm{p}, z \leftarrow w)) \mid \delta.$$

Consider two cases: either m is zero or m is $\mathrm{s}(p)$ for a canonical number p. In the former case, the value of the left-hand side is the value of $a \mid \gamma$ and that of the right-hand side is that of $a \circ \sigma \mid \delta$, and these two elements are equal. In the second case, it may be assumed that the two sides are equal with p instead of m; call these equal values e and f. According to (C5.37), the value of the left-hand side is the value of $b \mid (\gamma, y \leftarrow p, z \leftarrow e)$, say s; and that of the right-hand side is that of $b \circ (\sigma \circ \mathrm{p} \circ \mathrm{p}, y \leftarrow v \circ \mathrm{p}, z \leftarrow w) \mid (\delta, v \leftarrow p, w \leftarrow f)$, but, leaving out the type information,

$$\frac{\dfrac{\dfrac{\dfrac{\sigma \mid \delta \Rightarrow \gamma}{\sigma \circ \mathrm{p} \mid (\delta, v \leftarrow p) \Rightarrow \gamma}}{\sigma \circ \mathrm{p} \circ \mathrm{p} \mid (\delta, v \leftarrow p, w \leftarrow f) \Rightarrow \gamma} \quad \dfrac{v \mid (\delta, v \leftarrow p) \Rightarrow p}{v \circ \mathrm{p} \mid (\delta, v \leftarrow p, w \leftarrow f) \Rightarrow p}}{(\sigma \circ \mathrm{p} \circ \mathrm{p}, y \leftarrow v \circ \mathrm{p}) \mid (\delta, v \leftarrow p, w \leftarrow f) \Rightarrow (\gamma, y \leftarrow p)} \quad w \mid (\delta, v \leftarrow p, w \leftarrow f) \Rightarrow f}{(\sigma \circ \mathrm{p} \circ \mathrm{p}, y \leftarrow v \circ \mathrm{p}, z \leftarrow w) \mid (\delta, v \leftarrow p, w \leftarrow f) \Rightarrow (\gamma, y \leftarrow p, z \leftarrow f)} ,$$

and, since e and f are equal, the two sides are equal also in this case. This completes the justification.

To complete the treatment of the constant R, it remains to show that the computation rules are valid also under an assumption, i.e., it remains to justify the inference rules

$$\frac{\begin{array}{l} C : \mathrm{set}\ (\Gamma, x : \mathrm{el}(\mathrm{N})) \\ a : \mathrm{el}(C \circ (\mathrm{id}, x \leftarrow 0))\ (\Gamma) \\ b : \mathrm{el}(C \circ (\mathrm{p}, x \leftarrow \mathrm{s}(y)) \circ \mathrm{p})\ (\Gamma, y : \mathrm{el}(\mathrm{N}), z : \mathrm{el}(C \circ (\mathrm{p}, x \leftarrow y))) \end{array}}{\mathrm{R}_{\hat{x}C}(0, a, \hat{y}\hat{z}b) = a : \mathrm{el}(C \circ (\mathrm{id}, x \leftarrow 0))\ (\Gamma)} \tag{J5.82}$$

and

$$\frac{\begin{array}{l} p : \mathrm{el}(\mathrm{N})\ (\Gamma) \\ C : \mathrm{set}\ (\Gamma, x : \mathrm{el}(\mathrm{N})) \\ a : \mathrm{el}(C \circ (\mathrm{id}, x \leftarrow 0))\ (\Gamma) \\ b : \mathrm{el}(C \circ (\mathrm{p}, x \leftarrow \mathrm{s}(y)) \circ \mathrm{p})\ (\Gamma, y : \mathrm{el}(\mathrm{N}), z : \mathrm{el}(C \circ (\mathrm{p}, x \leftarrow y))) \end{array}}{\begin{array}{l} \mathrm{R}_{\hat{x}C}(a, \hat{y}\hat{z}b, \mathrm{s}(p)) = b \circ (\mathrm{id}, y \leftarrow p, z \leftarrow \mathrm{R}_{\hat{x}C}(p, a, \hat{y}\hat{z}b)) \\ \quad : \mathrm{el}(C \circ (\mathrm{id}, x \leftarrow \mathrm{s}(p)))\ (\Gamma) \end{array}} . \tag{J5.83}$$

These inference rules, and other rules of the same kind, will be called *reduction rules*.

Justification of (J5.82). Let the premisses be given and let γ be an arbitrarily given canonical assignment for Γ. The computation trace of the right-hand side is

$$\mathrm{el}(C \circ (\mathrm{id}, x \leftarrow 0) \mid \gamma) : a \mid \gamma \Rightarrow d \ \varepsilon\ \mathrm{el}(D),$$

and that of the left-hand side is

$$\frac{\mathrm{el}(\mathrm{N}) : 0 \mid \gamma \Rightarrow 0 \ \varepsilon\ \mathrm{el}(\mathrm{N}) \qquad \dfrac{\mathrm{el}(C \circ (\mathrm{id}, x \leftarrow 0) \mid \gamma) : a \mid \gamma \Rightarrow d \ \varepsilon\ \mathrm{el}(D)}{\mathrm{el}(C \mid (\gamma, x \leftarrow 0)) : \mathrm{R}_{\hat{x}C}[0](a, \hat{y}\hat{z}b) \mid \gamma \Rightarrow d \ \varepsilon\ \mathrm{el}(D)}}{\mathrm{el}(C \circ (\mathrm{id}, x \leftarrow 0) \mid \gamma) : \mathrm{R}_{\hat{x}C}(0, a, \hat{y}\hat{z}b) \Rightarrow d \ \varepsilon\ \mathrm{el}(D)} .$$

Since the result is the same in both cases, the two sides are equal.

Justification of (J5.83). Let the premisses be given and let γ be an arbitrarily given canonical assignment for Γ. Leaving out the type information for typographical reasons, the computation trace of the right-hand side is

$$\frac{\dfrac{\dfrac{\mathrm{id} \mid \gamma \Rightarrow \gamma \quad p \mid \gamma \Rightarrow q}{(\mathrm{id}, y \leftarrow p) \mid \gamma \Rightarrow (\gamma, y \leftarrow q)} \qquad \dfrac{p \mid \gamma \Rightarrow q \quad \mathrm{R}_{\hat{x}C}[q](a, \hat{y}\hat{z}b) \mid \gamma \Rightarrow e}{\mathrm{R}_{\hat{x}C}(p, a, \hat{y}\hat{z}b) \mid \gamma \Rightarrow e}}{(\mathrm{id}, y \leftarrow p, z \leftarrow \mathrm{R}_{\hat{x}C}(p, a, \hat{y}\hat{z}b)) \mid \gamma \Rightarrow (\gamma, x \leftarrow q, z \leftarrow e)} \qquad b \mid (\gamma, y \leftarrow q, z \leftarrow e) \Rightarrow d}{b \circ (\mathrm{id}, y \leftarrow p, z \leftarrow \mathrm{R}_{\hat{x}C}(p, a, \hat{y}\hat{z}b)) \mid \gamma \Rightarrow d}$$

and that of the left-hand side is

$$\frac{\dfrac{p \mid \gamma \Rightarrow q}{\mathrm{s}(p) \mid \gamma \Rightarrow \mathrm{s}(q)} \qquad \dfrac{\mathrm{R}_{\hat{x}C}[q](a, \hat{y}\hat{z}b) \mid \gamma \Rightarrow e \quad b \mid (\gamma, y \leftarrow q, z \leftarrow e) \Rightarrow d}{\mathrm{R}_{\hat{x}C}[\mathrm{s}(q)](a, \hat{y}\hat{z}b) \mid \gamma \Rightarrow d}}{\mathrm{R}_{\hat{x}C}(\mathrm{s}(p), a, \hat{y}\hat{z}b) \mid \gamma \Rightarrow d} .$$

The result is d in both cases. This completes the justification.

This completes the treatment of the constant R for recursion on the natural numbers.

It remains to treat of the elimination rules for the sets B, 1, $\emptyset$, $A \times B$, $A + B$, $\mathrm{L}(A, n)$, and $(\Sigma x : A)B$. Since $A \times B$ could have been defined as $(\Sigma x : A)B \circ \mathrm{p}$, just as $A \to B$ was defined as $(\Pi x : A)B \circ \mathrm{p}$, I will not give the elimination rule for $A \times B$ separately. For the remaining sets, I will only give the inference rules with some explanations, but without the detailed treatment given to the elimination rule for the natural numbers.

The finite sets B, 1, and $\emptyset$, with two, one, and zero elements, respectively, have the following inference rules: for the set B,

$$\frac{\begin{array}{ll} n : \mathrm{el}(\mathrm{B})\ (\Gamma) & a : \mathrm{el}(C \circ (\mathrm{id}, x \leftarrow 1))\ (\Gamma) \\ C : \mathrm{set}\ (\Gamma, x : \mathrm{el}(\mathrm{B})) & b : \mathrm{el}(C \circ (\mathrm{id}, x \leftarrow 0))\ (\Gamma) \end{array}}{\mathrm{R}^2_{\hat{x}C}(n, a, b) : \mathrm{el}(C \circ (\mathrm{id}, x \leftarrow n))\ (\Gamma)} ; \tag{R5.37}$$

for the set 1,

$$\frac{\begin{array}{l} n : \mathrm{el}(1)\ (\Gamma) \\ C : \mathrm{set}\ (\Gamma, x : \mathrm{el}(1)) \quad a : \mathrm{el}(C \circ (\mathrm{id}, x \leftarrow 0))\ (\Gamma) \end{array}}{\mathrm{R}^1_{\hat{x}C}(n, a) : \mathrm{el}(C \circ (\mathrm{id}, x \leftarrow n))\ (\Gamma)}\ ; \tag{R5.38}$$

and for the set $\emptyset$,

$$\frac{n : \mathrm{el}(\emptyset)\ (\Gamma) \quad C : \mathrm{set}\ (\Gamma, x : \mathrm{el}(\emptyset))}{\mathrm{R}^0_{\hat{x}C}(n) : \mathrm{el}(C \circ (\mathrm{id}, x \leftarrow n))\ (\Gamma)}\ . \tag{R5.39}$$

These inference rules are recognized in virtue of computation rules similar to those for R: given a canonical assignment γ for Γ, first $n \mid \gamma$ is computed, and then the computation continues according to its value. For R^2, there are two cases, for R^1 there is only one case, and for R^0 there are zero cases, i.e., R^0 is a limiting case.

The constant R^2 corresponds to the "if...then...else..." construct in functional programming languages. Instead of $\mathrm{R}^2_{\hat{x}C}(n, a, b)$, one could write

if n then a else b.

For obvious reasons, the constant R^1 is seldom used. It has no counterpart in normal functional programming.

The constant R^0 corresponds to abort in programming,[16] with the difference that, in intuitionistic type theory, uses of abort have to be legitimate; that is, to use R^0, or abort, one has to prove that the branch of the computation in question cannot be taken, by exhibiting an element of the empty set in it.

In the three forms R^2, R^1, and R^0, equals can be substituted for equals, and substitutions can be moved to the parts; no variables are bound, except in the family C, which is treated as in the case of the natural numbers.

For R^2, it is easy to justify the inference rules

$$\frac{\begin{array}{r} a : \mathrm{el}(C \circ (\mathrm{id}, x \leftarrow 1))\ (\Gamma) \\ C : \mathrm{set}\ (\Gamma, x : \mathrm{el}(\mathrm{B})) \quad b : \mathrm{el}(C \circ (\mathrm{id}, x \leftarrow 0))\ (\Gamma) \end{array}}{\mathrm{R}^2_{\hat{x}C}(1, a, b) = a : \mathrm{el}(C \circ (\mathrm{id}, x \leftarrow 1))\ (\Gamma)} \tag{J5.84}$$

and

$$\frac{\begin{array}{r} a : \mathrm{el}(C \circ (\mathrm{id}, x \leftarrow 1))\ (\Gamma) \\ C : \mathrm{set}\ (\Gamma, x : \mathrm{el}(\mathrm{B})) \quad b : \mathrm{el}(C \circ (\mathrm{id}, x \leftarrow 0))\ (\Gamma) \end{array}}{\mathrm{R}^2_{\hat{x}C}(0, a, b) = b : \mathrm{el}(C \circ (\mathrm{id}, x \leftarrow 0))\ (\Gamma)}\ , \tag{J5.85}$$

once the computation rules are given. Similarly, for R^1, it is easy to justify the inference rule

$$\frac{C : \mathrm{set}\ (\Gamma, x : \mathrm{el}(\mathrm{B})) \quad a : \mathrm{el}(C \circ (\mathrm{id}, x \leftarrow 0))\ (\Gamma)}{\mathrm{R}^1_{\hat{x}C}(0, a) = a : \mathrm{el}(C \circ (\mathrm{id}, x \leftarrow 0))\ (\Gamma)}\ . \tag{J5.86}$$

For obvious reasons, there is no similar inference rule for R^0.

Now to the disjoint union $A + B$ of two sets, the eliminatory constant of which is written D, and which has the elimination rule

[16] Cf. Martin-Löf, 'Constructive mathematics and computer programming', p. 157.

$$\frac{\begin{array}{l} n : \mathrm{el}(A+B)\ (\varGamma) \qquad C : \mathrm{set}\ (\varGamma, x : \mathrm{el}(A+B)) \\ a : \mathrm{el}(C \circ (\mathrm{p}, x \leftarrow \mathrm{i}(y)))\ (\varGamma, y : \mathrm{el}(A)) \\ b : \mathrm{el}(C \circ (\mathrm{p}, x \leftarrow \mathrm{j}(z)))\ (\varGamma, z : \mathrm{el}(B)) \end{array}}{\mathrm{D}_{A,B,\hat{x}C}(n, \hat{y}a, \hat{z}b) : \mathrm{el}(C \circ (\mathrm{id}, x \leftarrow n))\ (\varGamma)} \ . \tag{R5.40}$$

The reduction rules are given by

$$\frac{\begin{array}{l} p : \mathrm{el}(A)\ (\varGamma) \qquad C : \mathrm{set}\ (\varGamma, x : \mathrm{el}(A+B)) \\ a : \mathrm{el}(C \circ (\mathrm{p}, x \leftarrow \mathrm{i}(y)))\ (\varGamma, y : \mathrm{el}(A)) \\ b : \mathrm{el}(C \circ (\mathrm{p}, x \leftarrow \mathrm{j}(z)))\ (\varGamma, z : \mathrm{el}(B)) \end{array}}{\mathrm{D}_{A,B,\hat{x}C}(\mathrm{i}(p), \hat{y}a, \hat{z}b) = a \circ (\mathrm{id}, y \leftarrow p) : \mathrm{el}(C \circ (\mathrm{id}, x \leftarrow \mathrm{i}(p)))\ (\varGamma)} \tag{J5.87}$$

and

$$\frac{\begin{array}{l} q : \mathrm{el}(B)\ (\varGamma) \qquad C : \mathrm{set}\ (\varGamma, x : \mathrm{el}(A+B)) \\ a : \mathrm{el}(C \circ (\mathrm{p}, x \leftarrow \mathrm{i}(y)))\ (\varGamma, y : \mathrm{el}(A)) \\ b : \mathrm{el}(C \circ (\mathrm{p}, x \leftarrow \mathrm{j}(z)))\ (\varGamma, z : \mathrm{el}(B)) \end{array}}{\mathrm{D}_{A,B,\hat{x}C}(\mathrm{j}(q), \hat{y}a, \hat{z}b) = b \circ (\mathrm{id}, z \leftarrow q) : \mathrm{el}(C \circ (\mathrm{id}, x \leftarrow \mathrm{j}(q)))\ (\varGamma)} \ ; \tag{J5.88}$$

these reduction rules also indicate, implicitly, how $D(n, \hat{y}a, \hat{z}b) \mid \gamma$ is computed.

Similarly, the eliminatory constant of the disjoint union $(\Sigma\, x : A)B$ of a family of sets is written E, and to it corresponds the elimination rule

$$\frac{\begin{array}{l} n : \mathrm{el}((\Sigma\, x : A)B)\ (\varGamma) \qquad C : \mathrm{set}\ (\varGamma, y : \mathrm{el}((\Sigma\, x : A)B)) \\ a : \mathrm{el}(C \circ (\mathrm{p} \circ \mathrm{p}, y \leftarrow (z \circ \mathrm{p}\,, w)))\ (\varGamma, z : \mathrm{el}(A), w : \mathrm{el}(B \circ (\mathrm{p}, x \leftarrow z))) \end{array}}{\mathrm{E}_{A,\hat{x}B,\hat{y}C}(n, \hat{z}\hat{w}a) :\ \mathrm{el}(C \circ (\mathrm{id}, y \leftarrow n))\ (\varGamma)} \tag{R5.41}$$

and the reduction rule

$$\frac{\begin{array}{l} p : \mathrm{el}(A)\ (\varGamma) \qquad q : \mathrm{el}(B \circ (\mathrm{id}, x \leftarrow p))\ (\varGamma) \\ C : \mathrm{set}\ (\varGamma, y : \mathrm{el}((\Sigma\, x : A)B)) \\ a : \mathrm{el}(C \circ (\mathrm{p} \circ \mathrm{p}, y \leftarrow (z \circ \mathrm{p}\,, w)))\ (\varGamma, z : \mathrm{el}(A), w : \mathrm{el}(B \circ (\mathrm{p}, x \leftarrow z))) \end{array}}{\begin{array}{r} \mathrm{E}_{A,\hat{x}B,\hat{y}C}((p\,, q), \hat{z}\hat{w}a) = a \circ (\mathrm{id}, z \leftarrow p, w \leftarrow q) : \\ \mathrm{el}(C \circ (\mathrm{id}, y \leftarrow (p\,, q)))\ (\varGamma) \end{array}} \ . \tag{J5.89}$$

If $A \times B$ is defined as $(\Sigma\, x : A)B \circ \mathrm{p}$, as indicated above, then the eliminatory constant E can be used also for the Cartesian product. In this case, the above inference rule can be simplified somewhat, since $B \circ \mathrm{p} \circ (\mathrm{id}, x \leftarrow p)$ is equal to B.

Finally, the set of lists over a set A of a certain length m, $\mathrm{L}(A, m)$, has no standard name for its eliminatory constant: for brevity of notation, I will simply write it S (the third letter of "list" and the last letter of "cons"). In the previous inference rules, weakening substitutions were inserted to make sure that the variables x, y, etc., used in these inference rules could be the same: in the following inference rule, these weakening substitutions will be omitted for brevity of notation; furthermore, I will write p^2 for $\mathrm{p} \circ \mathrm{p}$, etc., also for brevity of notation.

$$\frac{\begin{array}{l} n : \mathrm{el}(\mathrm{L}(A, m))\ (\Gamma) \qquad C : \mathrm{set}\ (\Gamma, x : \mathrm{el}(\mathrm{N}), y : \mathrm{el}(\mathrm{L}(A \circ \mathrm{p}, x))) \\ a : \mathrm{el}(C \circ (\mathrm{id}, x \leftarrow 0, y \leftarrow ()))\ (\Gamma) \\ b : \mathrm{el}(C \circ (\mathrm{p}^2, x \leftarrow \mathrm{s}(x), y \leftarrow (y, z)) \circ \mathrm{p})\ (\Gamma, x : \mathrm{el}(\mathrm{N}), \\ \quad y : \mathrm{el}(A \circ \mathrm{p}), z : \mathrm{el}(\mathrm{L}(A \circ \mathrm{p}^2, x)), w : \mathrm{el}(C \circ (\mathrm{p}^3, x \leftarrow x, y \leftarrow z))) \end{array}}{\mathrm{S}_{A,m,\hat{x}C}(n, a, \hat{y}\hat{z}b) : \mathrm{el}(C \circ (\mathrm{id}, x \leftarrow m, y \leftarrow n))\ (\Gamma)} \; . \qquad \text{(R5.42)}$$

This inference rule shows that a type-theoretic notation, like the present one, becomes embarrassingly cumbersome for more complicated inference rules.

The reduction rules for S are given by

$$\frac{\begin{array}{l} C : \mathrm{set}\ (\Gamma, x : \mathrm{el}(\mathrm{N}), y : \mathrm{el}(\mathrm{L}(A \circ \mathrm{p}, x))) \\ a : \mathrm{el}(C \circ (\mathrm{id}, x \leftarrow 0, y \leftarrow ()))\ (\Gamma) \\ b : \mathrm{el}(C \circ (\mathrm{p}^2, x \leftarrow \mathrm{s}(x), y \leftarrow (y, z)) \circ \mathrm{p})\ (\Gamma, x : \mathrm{el}(\mathrm{N}), \\ \quad y : \mathrm{el}(A \circ \mathrm{p}), z : \mathrm{el}(\mathrm{L}(A \circ \mathrm{p}^2, x)), w : \mathrm{el}(C \circ (\mathrm{p}^3, x \leftarrow x, y \leftarrow z))) \end{array}}{\mathrm{S}_{A,0,\hat{x}C}((), a, \hat{y}\hat{z}b) = a : \mathrm{el}(C \circ (\mathrm{id}, x \leftarrow 0, y \leftarrow ()))\ (\Gamma)} \qquad \text{(J5.90)}$$

and

$$\frac{\begin{array}{l} c : \mathrm{el}(A)\ (\Gamma) \qquad p : \mathrm{el}(\mathrm{L}(A, q))\ (\Gamma) \\ C : \mathrm{set}\ (\Gamma, x : \mathrm{el}(\mathrm{N}), y : \mathrm{el}(\mathrm{L}(A \circ \mathrm{p}, x))) \\ a : \mathrm{el}(C \circ (\mathrm{id}, x \leftarrow 0, y \leftarrow ()))\ (\Gamma) \\ b : \mathrm{el}(C \circ (\mathrm{p}^2, x \leftarrow \mathrm{s}(x), y \leftarrow (y, z)) \circ \mathrm{p})\ (\Gamma, x : \mathrm{el}(\mathrm{N}), \\ \quad y : \mathrm{el}(A \circ \mathrm{p}), z : \mathrm{el}(\mathrm{L}(A \circ \mathrm{p}^2, x)), w : \mathrm{el}(C \circ (\mathrm{p}^3, x \leftarrow x, y \leftarrow z))) \end{array}}{\begin{array}{l} \mathrm{S}_{A,\mathrm{s}(q),\hat{x}C}((c, p), a, \hat{y}\hat{z}b) = \\ b \circ (\mathrm{id}, x \leftarrow q, y \leftarrow c, z \leftarrow p, w \leftarrow \mathrm{S}_{A,q,\hat{x}C}(p, a, \hat{y}\hat{z}b)) : \\ \quad \mathrm{el}(C \circ (\mathrm{id}, x \leftarrow \mathrm{s}(q), y \leftarrow (c, p)))\ (\Gamma) \end{array}} \; . \qquad \text{(J5.91)}$$

§ 8. Propositions as sets

Propositions and their causes differ from sets and their elements only in the vocabulary used to speak about them, as indicated in Table 7. This surprising correspondence between intuitionistic logic—as presented in Chapter II, Section 7—and the intuitionistic theory of sets—presented in Chapters III and IV—, is called the Curry-Howard correspondence, after its discoverers.[17]

This correspondence was first seen as a merely formal correspondence between the inference rules of intuitionistic logic and the inference rules of combinatory logic. Now, combinatory logic, and its equivalent alternative, λ-calculus, can be viewed as a subsystem of the theory of sets given above;[18] thus, in view of the BHK interpretation of the logical connectives, the Curry-Howard correspondence is not surprising.

Though there are different *nuances* to the pairs proposition/cause and set/element, they nevertheless have exactly the same meaning in the narrow type-theoretic sense: a proposition is defined by laying down what counts as a

[17]Curry, 'Functionality in Combinatory Logic'; and Howard, 'The formulae-as-types notion of construction'.

[18]Cf. also Martin-Löf, *Intuitionistic Type Theory*, pp. 35–38.

logic	set theory/computer science
proposition	set/data set
cause	element/data element
true	inhabited
conjunction (&)	product (×)
disjunction ($\vee$)	sum (+)
implication ($\supset$)	function space ($\to$)
falsum (Λ)	empty set ($\emptyset$)
universal quantification ($\forall$)	product of a family of sets (Π)
existential quantification ($\exists$)	sum of a family of sets (Σ)
modus ponendo ponens	function application (app)
implication introduction	lambda abstraction (λ)
ex falso quodlibet	aborting a computation (R^0)
etc.	etc.

Table 7. The Curry-Howard correspondence between propositions and sets, causes and elements, etc.

cause of it and a set is defined by laying down what counts as an element of it. That is, the three bidirectional inference rules

$$\frac{A\ \varepsilon\ \text{set}}{A\ \varepsilon\ \text{prop}},\qquad \frac{A:\text{set}}{A:\text{prop}},\ \text{and}\qquad \frac{A:\text{set}\ (\Gamma)}{A:\text{prop}\ (\Gamma)}$$

can be taken as meaning determining for the forms of assertion of their conclusions; similarly, the inference rules

$$\frac{a\ \varepsilon\ \text{el}(A)}{a\ \varepsilon\ \text{cause}(A)},\qquad \frac{a:\text{el}(A)}{a:\text{cause}(A)},\ \text{and}\qquad \frac{a:\text{el}(A)\ (\Gamma)}{a:\text{cause}(A)\ (\Gamma)}$$

are meaning determining for their conclusions, i.e., *cause* of a proposition means the same as *element* of a set. These definitions of the notions of proposition and cause should be understood as clarifications of the definitions given in Chapter II, Section 6. As laid down in that section, that a proposition is true means that it has a cause, i.e., the inference rules

$$\frac{a\ \varepsilon\ \text{cause}(A)}{A\ \text{true}},\qquad \frac{a:\text{cause}(A)}{A\ \text{true}},\ \text{and}\qquad \frac{a:\text{cause}(A)\ (\Gamma)}{A\ \text{true}\ \ (\Gamma)}.$$

are meaning determining for the notion of truth. That a proposition is true has as its counterpart that a set has an element, i.e., that it is *inhabited*.[19]

The logical connectives &, $\vee$, $\supset$, and Λ, introduced in Chapter II, Section 7, can now be redefined by the nominal definitions (cf. Table 7):

$$A \,\&\, B \overset{\text{def}}{=} A \times B : \text{set},$$

$$A \vee B \overset{\text{def}}{=} A + B : \text{set},$$

$$A \supset B \overset{\text{def}}{=} A \to B : \text{set},\ \text{and}$$

$$\Lambda \overset{\text{def}}{=} \emptyset : \text{set}.$$

[19]At this point, one has to be careful with the terminology, since to say that a set A is *nonempty* could also be taken to mean that it is not the case that A is empty.

We can now go back to the inference rules of Chapter II, Section 7, and see that they are possible to justify with explicit causes, i.e., with names assigned to the causes. For example, the application rule (R5.31), formulated in terms of causes reads

$$\frac{f : \text{cause}(A \supset B)\ (\Gamma) \quad a : \text{cause}(A)\ (\Gamma)}{\text{app}(f, a) : \text{cause}(B)\ (\Gamma)}\ ;$$

if the explicit causes are replaced by truth, the result is the rule of *modus ponendo ponens*:

$$\frac{A \supset B \text{ true }\ (\Gamma) \quad A \text{ true }\ (\Gamma)}{B \text{ true }\ (\Gamma)}\ .$$

Similarly, *modus ponendo tollens*

$$\frac{{\sim}(A \mathbin{\&} B) \text{ true }\ (\Gamma) \quad A \text{ true }\ (\Gamma)}{{\sim}B \text{ true }\ (\Gamma)}$$

becomes

$$\frac{f : \text{cause}((A \mathbin{\&} B) \supset \Lambda)\ (\Gamma) \quad a : \text{cause}(A)\ (\Gamma)}{(\lambda\, x)\, \text{app}(f \circ \text{p}, (a \circ \text{p}, x)) : \text{cause}(B \supset \Lambda)\ (\Gamma)}$$

when the causes are filled in. An inference rule for which it is particularly illuminating to make the causes explicit is *ex falso quodlibet*, i.e.,

$$\frac{\Lambda \text{ true }\ (\Gamma) \quad (A : \text{prop}\ (\Gamma))}{A \text{ true }\ (\Gamma)}\ ,$$

which becomes

$$\frac{n : \text{cause}(\Lambda)\ (\Gamma) \quad A : \text{prop}\ (\Gamma)}{\text{R}^0_{\hat{x}A\circ\text{p}}(n) : \text{cause}(A)\ (\Gamma)}$$

after (R5.39); note that $A \circ \text{p} \circ (\text{id}, x \leftarrow n) = A : \text{prop}\ (\Gamma)$.

The BHK interpretation can now be extended from propositional logic to predicate logic through the nominal definitions

$$(\exists x : A)B \overset{\text{def}}{=} (\Sigma\, x : A)B$$

and

$$(\forall x : A)B \overset{\text{def}}{=} (\Pi\, x : A)B.$$

Here A is understood as a set and B as a propositional function on A; that is, even though proposition and set can be used interchangeably, the two quantifiers are understood as having the formation rules

$$\frac{A : \text{set}\ (\Gamma) \quad B : \text{prop}\ (\Gamma, x : \text{el}(A))}{(\forall x : A)B : \text{prop}\ (\Gamma)}$$

and

$$\frac{A : \text{set}\ (\Gamma) \quad B : \text{prop}\ (\Gamma, x : \text{el}(A))}{(\exists x : A)B : \text{prop}\ (\Gamma)}\ ,$$

cf. (R5.28) and (R5.33). The standard laws governing the quantifiers in predicate logic are the result of suppressing the causes in the corresponding inference rules for the sets Π and Σ: the rule of $\forall$-introduction

$$\frac{B \text{ true }\ (\Gamma, x : \text{el}(A))}{(\forall x : A)B \text{ true }\ (\Gamma)}$$

comes from (R5.30); the rule of ∀-elimination

$$\frac{(\forall x : A)B \text{ true } \ (\Gamma) \quad a : \text{el}(A)\ (\Gamma)}{B \circ (\text{id}, x \leftarrow a) \text{ true } \ (\Gamma)}$$

comes from (R5.31); the rule of ∃-introduction

$$\frac{a : \text{el}(A)\ (\Gamma) \quad (B : \text{prop}\ (\Gamma, x : \text{el}(A))) \quad B \circ (\text{id}, x \leftarrow a) \text{ true } \ (\Gamma)}{(\exists x : A)B \text{ true } \ (\Gamma)}$$

comes from (R5.34); and the rule of ∃-elimination

$$\frac{(\exists x : A)B \text{ true } \ (\Gamma) \quad D \circ \text{p} \circ \text{p true } \ (\Gamma, x : \text{el}(A), y : \text{cause}(B))}{D \text{ true } \ (\Gamma)}$$

comes from (R5.41). Note that (R5.41) reads

$$\frac{\begin{array}{l} n : \text{cause}((\exists x : A)B)\ (\Gamma) \\ C : \text{prop}\ (\Gamma, y : \text{cause}((\exists x : A)B)) \\ a : \text{cause}(C \circ (\text{p} \circ \text{p}, y \leftarrow (x, z)))\ (\Gamma, x : \text{el}(A), z : \text{cause}(B)) \end{array}}{\text{E}_{A, \hat{x}B, \hat{y}C}(n, a) : \text{cause}(C \circ (\text{id}, y \leftarrow n))\ (\Gamma)}$$

in terms of propositions and causes; to justify ∃-elimination, take C to be

$$C = D \circ \text{p} : \text{prop}\ (\Gamma, y : \text{cause}((\exists x : A)B)),$$

and note that

$$D \circ \text{p} \circ (\text{p} \circ \text{p}, y \leftarrow (x, z)) = D \circ \text{p} \circ \text{p} : \text{prop}\ (\Gamma, x : \text{el}(A), z : \text{cause}(B)).$$

The rule of ∃-elimination, which *is* possible to formulate without mentioning causes, is significantly weaker than (R5.41), which *cannot* be formulated without mentioning causes.[20]

[20]Cf. Martin-Löf, *Intuitionistic Type Theory*, pp. 39-52.

CHAPTER VI

Intuitionism

IT WAS RECOGNIZED already by the authors of *Principia Mathematica* that using the law of excluded middle, or its equivalent, proof by contradiction, to *prove* the law of excluded middle involves a vicious circle.[1] In view of this, it is astonishing that the critics of Brouwer's rejection of the law of excluded middle claimed that his rejection leads to a third truth value, which is inconsistent,[2] and that Church had to correct his fellow logicians by restating that their argument involves a vicious circle.[3] The first section of this chapter is concerned with the intuitionistic interpretation of proofs by contradiction, i.e., of apagogical proofs. This analysis will put the intuitionistic rejection of the law of excluded middle in perspective. The second section treats of some philosophical and metaphysical aspects of the law of excluded middle. The third section consists of a critique of formalism and set-theoretical Platonism as approaches to the foundations of mathematics.

§ 1. The intuitionistic interpretation of apagoge

The laws of intuitionistic propositional logic were demonstrated in Chapter II, Section 7. To deny the equivalence of the propositions A and $\sim\sim A$ is a bold step to take but, I think, a necessary one. However, two things should be noted. First, that A implies $\sim\sim A$ for any proposition A, i.e., that the inference rule

$$\frac{A : \text{prop}}{A \supset \sim\sim A \text{ true}}$$

is valid, as demonstrated by

$$\cfrac{\cfrac{\cfrac{\cfrac{A : \text{prop}}{\sim A \text{ true} \ \ (A \text{ true}, \sim A \text{ true})} \quad \cfrac{A : \text{prop}}{A \text{ true} \ \ (A \text{ true}, \sim A \text{ true})}}{\Lambda \text{ true} \ \ (A \text{ true}, \sim A \text{ true})}}{\sim\sim A \text{ true} \ \ (A \text{ true})}}{A \supset \sim\sim A \text{ true}} .$$

Second, that the two propositions $\sim A$ and $\sim\sim\sim A$ are equivalent.[4] One part of this equivalence is a special case of the law established above, and the other

[1]Whitehead and Russell, *Principia Mathematica*, Intro., Ch. 2, § 1, p. 40.
[2]Cf. Mancosu, *From Brouwer to Hilbert*, pp. 278–280.
[3]Church, 'On the law of excluded middle', p. 77.
[4]This was first demonstrated by Brouwer, 'Intuitionistische Zerlegung mathematischer Grundbegriffe', p. 253.

J.G. Granström, *Treatise on Intuitionistic Type Theory*, Logic, Epistemology, and the Unity of Science 22, DOI 10.1007/978-94-007-1736-7_6,

part is given by the inference rule

$$\frac{A : \text{prop}}{{\sim}{\sim}{\sim}A \supset {\sim}A \text{ true}}\,,$$

which is demonstrated by[5]

$$\dfrac{\dfrac{\dfrac{A : \text{prop}}{{\sim}{\sim}{\sim}A \text{ true}\ \ ({\sim}{\sim}{\sim}A \text{ true})} \qquad \dfrac{\dfrac{A : \text{prop}}{A \supset {\sim}{\sim}A \text{ true}} \quad \dfrac{A : \text{prop}}{A \text{ true}\ \ (A \text{ true})}}{{\sim}{\sim}A \text{ true}\ \ (A \text{ true})}}{\Lambda \text{ true}\ \ ({\sim}{\sim}{\sim}A \text{ true}, A \text{ true})}}{\dfrac{{\sim}A \text{ true}\ \ ({\sim}{\sim}{\sim}A \text{ true})}{{\sim}{\sim}{\sim}A \supset {\sim}A \text{ true}}}\,.$$

Thus, negative propositions are equivalent to their double negation, but positive propositions need not be. Instead of *duplex negatio affirmat*, intuitionistic logic has *triplex negatio negat*.

Keeping these logical laws in mind, I will now investigate the distinction between the two assertions

$$A \text{ true}$$

and

$${\sim}{\sim}A \text{ true}$$

in greater detail.

A distinction made by Aristotle in connection with syllogistic reasoning is between *direct* proof and *indirect* proof (proof *per impossibile*).[6] A direct proof proceeds by inference rules, as we are used to. In an indirect proof of A, one assumes the negation of A and shows that this assumption leads to a contradiction: with the intuitionistic interpretation of negation, this leads to an intuitionistic proof of ${\sim}{\sim}A$. The distinction between direct and indirect proofs was upheld by Kant, using the Greek words ostensive and apagogical.

"The third rule peculiar to pure reason, in so far as it is to be subjected to a discipline in respect of transcendental proofs, is that its proofs must never be *apagogical*, but always *ostensive*. The direct or ostensive proof, in every kind of knowledge, is that which combines with the conviction of its truth insight into the sources of its truth; the apagogical proof, on the other hand, while it can indeed yield certainty, cannot enable us to comprehend truth in its connection with the grounds of its possibility. The latter is therefore to be regarded rather as a last resort than as a mode of procedure which satisfies all the requirements of reason."[7]

In the history of logic, there is also another topic of importance to the distinction between A being true and ${\sim}{\sim}A$ being true, namely, the topic of

[5]In this demonstration, and in what follows, weakening is implicit, i.e., applications of the weakening rule are not written out. Moreover, when making an assumption, the demonstration of the well-formedness of the assumption is left out.

[6]Cf. Aristotle, *An. Pr.*, Bk. 2, Ch. 14.

[7]Kant, *Kritik der reinen Vernunft*, Pt. 2.1.4, p. 513 (B 817) (trans. N. K. Smith).

causal proofs.[8] In brief: Aristotle made a distinction between demonstration of a fact (ὅτι) and demonstration of the reason for it (διότι).[9] In Latin, these terms were rendered *quia* and *propter quid*, i.e., demonstration *that* and demonstration *because of something*. Next, Averroës developed this distinction further by adding a third kind of demonstration, *potissima*, i.e., best of all, which is a simultaneous demonstration of the fact and the reason for it.[10] This distinction is called for if one admits inductive (or, better, abductive) reasoning from effect to cause, which then would be *propter quid* but not of a fact, because the conclusion is not necessary. Since such demonstrations are not accepted in mathematics, I will make no further use of this distinction but instead consider *propter quid* and *potissima* as synonymous. During the Renaissance, some authors claimed that there are no causes in mathematics, so its demonstrations cannot be *potissima*;[11] Biancani, among others, replied that the demonstrations of mathematics are *potissima* since they are by formal or material cause.[12] Indirect proofs were generally not considered causal.[13] Now the distinction became that between proofs that proceed by causes (*potissima*) and proofs that do not (*quia*): the former yield evidence while the latter only yield certainty. That is, something is certain if it cannot be otherwise and evident if known by its causes:

"Archimedes' admirers need to excuse his oblique procedure; both because it is long and complicated in the constructions and the proofs and because it is not completely satisfactory, since it produces certainty but not evidence. I am of the opinion that everything evident is certain but not everything certain is evident."[14]

It is natural to identify a proposition A being *evident* in Nardi's sense with it being *true* in our sense, and a proposition being *certain* in Nardi's sense with $\sim\sim A$ being true; because A implies $\sim\sim A$ but not the other way around, i.e., "everything evident is certain but not everything certain is evident". Through this long and, admittedly, inconclusive line of argument, Aristotle's distinction between *quia* and *propter quid* is reduced to that between $\sim\sim A$ being true and A being true.

The use of the word *certain* is somewhat unfortunate in this context because it suggests some kind of epistemic modality. It is clear that the use of the word *certain* in Nardi's distinction between kinds of evidence established by different mathematical proofs or constructions due to Archimedes is not the same as the use of the word in the distinction between knowledge and certainty.[15] Rather, certainty in Nardi's sense is a kind of knowledge—but of

[8]For a comprehensive treatment of this topic, the reader is referred to the first two chapters of Mancosu's book *Philosophy of Mathematics and Mathematical Practice in the Seventeenth Century*.
[9]Aristotle, *An. Post.*, Bk. 1, Ch. 13; and ibid., Bk. 2, Ch. 1.
[10]Mancosu, *Philosophy of Mathematics and Mathematical Practice in the Seventeenth Century*, p. 12.
[11]Ibid., p. 13.
[12]Ibid., p. 17.
[13]Ibid., p. 25.
[14]Nardi, quoted in ibid., p. 63.
[15]As discussed in Moore's 1941 Howison lecture 'Certainty' and Wittgenstein's book *On Certainty*.

what? We have already identified the sentence "A is evident" with the sentence "I know A is true" and the sentence "A is certain" with the sentence "I know $\sim\sim A$ is true"; but instead of saying that $\sim\sim A$ is true, we can say that A is *irrefutable*, i.e., that its negation does not admit a proof. Now *true* and *irrefutable* are, as it were, on the same level: we can know that A is true and we can know that A is irrefutable.

In his introduction to intuitionism, Heyting makes use of the distinction between negation *de jure* and negation *de facto*:[16] the former is the intuitionistic negation, while the latter negation has the property that $\sim\sim A$ entails A. This distinction becomes clearer if we identify *de jure* negation with the negation of the proposition A in the assertion that A is evident, or true, and *de facto* negation with the negation of A in the assertion that A is irrefutable:[17] with this distinction, both negations are the ordinary intuitionistic negation, but if A is negated twice in the assertion that A is irrefutable, we get that $\sim\sim A$ is irrefutable, or, which amounts to the same, that $\sim\sim\sim\sim A$ is true, which entails that A is irrefutable. Thus, the terms *de jure* and *de facto* could instead be applied to the proposition A, just as evident and certain, i.e., that A *de facto* is true, or that A is a fact, can be taken to mean that $\sim\sim A$ is true, i.e., that A is irrefutable.

Finally, Bolzano revived the Aristotelian distinction between *quia* and *propter quid* and made a distinction between Gewissmachungen and Begründungen, i.e., certifications and groundings.[18] For Bolzano, this distinction is not the same as that between apagogical and ostensive, but, again, a lot of what is said about the difference between certifications and groundings makes sense when a certification is taken to be a demonstration of $\sim\sim A$ being true and a grounding a demonstration of A being true.

Thus, I think that the essence of the observations which lead the various authors to make these distinctions really is that between $\sim\sim A$ being true and A being true, but, as expected, not everything written on the matter supports this.

I will use the abbreviation 'A false' for $\sim A$ being true, and the abbreviation 'A irrefutable' for $\sim\sim A$ being true, or, which amounts to the same, for $\sim A$ being false. That is, the bidirectional inference rules

$$\frac{\sim A \text{ true}}{\overline{A \text{ false}}} \quad \text{and} \quad \frac{\sim\sim A \text{ true}}{\overline{A \text{ irrefutable}}}$$

are valid. Combining them, we also get

$$\frac{\sim A \text{ false}}{\overline{A \text{ irrefutable}}} .$$

As demonstrated above, every true proposition is irrefutable, i.e., we have the inference rule

$$\frac{A \text{ true}}{A \text{ irrefutable}} .$$

[16] Heyting, *Intuitionism: An Introduction*, p. 18.
[17] Cf. ibid., Th. 1, p. 17.
[18] Sebestik, 'Bolzano's Logic'.

Moreover, irrefutability and truth coincide for negative propositions, i.e., we have the bidirectional inference rule

$$\frac{\sim A \text{ irrefutable}}{\sim A \text{ true}} \ .$$

Observe also that irrefutability and truth coincide for *falsum*, i.e., we have the bidirectional inference rule

$$\frac{\Lambda \text{ irrefutable}}{\Lambda \text{ true}} \ .$$

This becomes clear upon remembering that the proposition Λ being irrefutable is tantamount to the proposition $(\Lambda \supset \Lambda) \supset \Lambda$ being true (which, indeed, it is not).

The principle of proof by contradiction,[19] can now be formulated as the following special case of implication introduction[20]

$$\frac{\Lambda \text{ true} \quad (A \text{ false})}{A \text{ irrefutable}} \ .$$

That the conclusion of this inference rule is that A is irrefutable fits well with the view that proofs *per impossibile* do not give causal knowledge of the conclusion.[21]

Using the notation A false instead of $\sim A$ true, the inference rule *modus tollendo tollens* can be reformulated as

$$\frac{A \supset B \text{ true} \quad B \text{ false}}{A \text{ false}} \ ,$$

the inference rule *modus ponendo tollens* as

$$\frac{A \,\&\, B \text{ false} \quad A \text{ true}}{B \text{ false}} \ ,$$

and the inference rule *modus tollendo ponens* becomes

$$\frac{A \vee B \text{ true} \quad A \text{ false}}{B \text{ true}} \ .$$

Moreover, the principle of noncontradiction can be reformulated as

$$\frac{A \text{ false} \quad A \text{ true}}{\Lambda \text{ true}} \ .$$

We now have two distinct notions: A true and A irrefutable. Intuitionistic logic is primarily concerned with what is true, i.e., evident or *per causas*. It remains to show that the laws of logic are valid also when dealing with

---

[19]Also called proof *per contradictionem* or *per impossibile*, *reductio ad absurdum* or *ad impossibile*.

[20]In this inference rule, I have taken the liberty to use A false as an assumption. This is to be understood as tantamount to making the assumption $\sim A$ true. Similarly, an assumption of the form A irrefutable is tantamount to an assumption of the form $\sim\sim A$ true.

[21]Mancosu, *Philosophy of Mathematics and Mathematical Practice in the Seventeenth Century*, p. 25.

apagogical knowledge, or knowledge of irrefutable propositions.[22] Any inference rule of the general form

$$\frac{A_1 \text{ true} \quad \cdots \quad A_n \text{ true}}{B \text{ true}}$$

has a corresponding mediate inference rule

$$\frac{A_1 \text{ irrefutable} \quad \cdots \quad A_n \text{ irrefutable}}{B \text{ irrefutable}} .$$

That is, we can reason from irrefutable premisses to an irrefutable conclusion in exactly the same way as we reason from true premisses to a true conclusion, and, in addition, use the principle of proof by contradiction when demonstrating the irrefutability of the conclusion.

The above observation can be divided into two parts. First, if an implication $A \supset B$ is true, then the corresponding doubly negated implication $\sim\sim A \supset \sim\sim B$ is also true, i.e., the inference rule

$$\frac{A \supset B \text{ true}}{\sim\sim A \supset \sim\sim B \text{ true}}$$

is valid. The easiest way to convince oneself that this law is valid is to view it as two successive applications of the inference rule

$$\frac{A \supset B \text{ true}}{\sim B \supset \sim A \text{ true}} ,$$

demonstrated by *modus tollendo tollens*:

$$\cfrac{\cfrac{A \supset B \text{ true} \quad \cfrac{B : \text{prop}}{B \text{ false} \;\; (B \text{ false})}}{A \text{ false} \;\; (B \text{ false})}}{\sim B \supset \sim A \text{ true}} .$$

For each application of this inference rule, the antecedent and consequent are interchanged and negated; applying it twice to $A \supset B$ gives $\sim\sim A \supset \sim\sim B$. Next, the double negation of a conjunction is equivalent to the double negation of the conjuncts, i.e., the inference rules

$$\frac{A \,\&\, B \text{ irrefutable}}{A \text{ irrefutable}} , \qquad \frac{A \,\&\, B \text{ irrefutable}}{B \text{ irrefutable}} ,$$

and

$$\frac{A \text{ irrefutable} \quad B \text{ irrefutable}}{A \,\&\, B \text{ irrefutable}}$$

[22]The following considerations have a metamathematical counterpart in the double negation interpretation, first presented by Kolmogorov ('On the principle of excluded middle'). Cf. also Glivenko, 'Sur quelques points de la logique de M. Brouwer'; Gödel, 'Zur intuitionistische Arithmetik und Zahlentheorie'; and Gentzen, 'Die Widerspruchfreiheit der reinen Zahlentheorie'.

are valid. The first inference rule is demonstrated by

$$\dfrac{\dfrac{A \,\&\, B \text{ irrefutable} \qquad \dfrac{\dfrac{\dfrac{A : \text{prop}}{A \text{ false } (A \text{ false})} \qquad \dfrac{\dfrac{A : \text{prop} \quad B : \text{prop}}{A \,\&\, B \text{ true } (A \,\&\, B \text{ true})}}{A \text{ true } (A \,\&\, B \text{ true})}}{\Lambda \text{ true } (A \text{ false}, A \,\&\, B \text{ true})}}{A \,\&\, B \text{ false } (A \text{ false})}}{\Lambda \text{ true } (A \text{ false})}}{A \text{ irrefutable}} .$$

Similarly, *mutatis mutandis*, one can conclude that B is irrefutable from $A \,\&\, B$ being irrefutable. The third inference rule, with the conclusion that $A \,\&\, B$ is irrefutable, is demonstrated by

$$\dfrac{\dfrac{A \text{ irrefutable} \qquad \dfrac{\dfrac{B \text{ irrefutable} \qquad \dfrac{\dfrac{A : \text{prop} \quad B : \text{prop}}{A \,\&\, B \text{ false } (A \,\&\, B \text{ false})} \qquad \dfrac{A : \text{prop}}{A \text{ true } (A \text{ true})}}{B \text{ false } (A \,\&\, B \text{ false}, A \text{ true})}}{\Lambda \text{ true } (A \,\&\, B \text{ false}, A \text{ true})}}{A \text{ false } (A \,\&\, B \text{ false})}}{\Lambda \text{ true } (A \,\&\, B \text{ false})}}{A \,\&\, B \text{ irrefutable}} .$$

Combining the above, it becomes clear that we can reason from irrefutable premisses to an irrefutable conclusion in the same way as we reason from true premisses to a true conclusion; because if B can be inferred from A_1 up to A_n, then the implication $(A_1 \,\&\, \cdots \,\&\, A_n) \supset B$ is true, which entails that the implication $\sim\sim(A_1 \,\&\, \cdots \,\&\, A_n) \supset \sim\sim B$ is true as well, but $\sim\sim(A_1 \,\&\, \cdots \,\&\, A_n)$ is equivalent to $\sim\sim A_1 \,\&\, \cdots \,\&\, \sim\sim A_n$, so the implication $\sim\sim A_1 \,\&\, \cdots \,\&\, \sim\sim A_n \supset \sim\sim B$ is true, whence, by *modus ponendo ponens*, the inference with $\sim\sim A_1$ up to $\sim\sim A_n$ as premisses and $\sim\sim B$ as conclusion is valid.

An inference rule which looks surprising at first sight, but which nevertheless can be demonstrated, is

$$\frac{B \text{ irrefutable } (A \text{ true})}{A \supset B \text{ irrefutable}} .$$

That is, when proving that an implication is irrefutable, we can make the strong assumption that the antecedent is true, instead of irrefutable. To demonstrate this, we need to make use of an assumption of the form $A \supset B$ false. The inference rules

$$\frac{A \supset B \text{ false}}{A \text{ irrefutable}}$$

and

$$\frac{A \supset B \text{ false}}{B \text{ false}}$$

are both valid, as demonstrated by

$$\dfrac{\dfrac{A \supset B \text{ false} \qquad \dfrac{\dfrac{\dfrac{\dfrac{A : \text{prop}}{A \text{ false } \ (A \text{ false})} \quad \dfrac{A : \text{prop}}{A \text{ true } \ (A \text{ true})}}{\Lambda \text{ true } \ (A \text{ false}, A \text{ true})}}{B \text{ true } \ (A \text{ false}, A \text{ true})}}{A \supset B \text{ true } \ (A \text{ false})}}{\Lambda \text{ true } \ (A \text{ false})}}{A \text{ irrefutable}}$$

and

$$\dfrac{\dfrac{A \supset B \text{ false} \qquad \dfrac{\dfrac{A : \text{prop} \quad \dfrac{B : \text{prop}}{B \text{ true } \ (B \text{ true})}}{B \text{ true } \ (B \text{ true}, A \text{ true})}}{A \supset B \text{ true } \ (B \text{ true})}}{\Lambda \text{ true } \ (B \text{ true})}}{B \text{ false}} \, .$$

The idea used in the last demonstration is that, if B is true, then $A \supset B$ is also true. Using these inference rules, we get

$$\dfrac{\dfrac{\dfrac{\dfrac{A : \text{prop} \quad B : \text{prop}}{A \supset B \text{ false } (A \supset B \text{ false})}}{A \text{ irrefutable } (A \supset B \text{ false})} \qquad \dfrac{\dfrac{B \text{ irref.} \, (A \text{ true}) \quad \dfrac{\dfrac{A : \text{prop} \quad B : \text{prop}}{A \supset B \text{ false } (A \supset B \text{ false})}}{B \text{ false } \ (A \supset B \text{ false})}}{\Lambda \text{ true } \ (A \supset B \text{ false}, A \text{ true})}}{A \text{ false } \ (A \supset B \text{ false})}}{\Lambda \text{ true } \ (A \supset B \text{ false})}}{A \supset B \text{ irrefutable}} \, .$$

Finally, a well-known result in intuitionistic logic is that every proposition of the form $A \vee \sim A$ is irrefutable,[23] i.e., that we have the inference rule

$$\dfrac{A : \text{prop}}{A \vee \sim A \text{ irrefutable}} \, .$$

To demonstrate this logical law, note that the inference rules

$$\dfrac{A \vee B \text{ false}}{A \text{ false}} \qquad \text{and} \qquad \dfrac{A \vee B \text{ false}}{B \text{ false}}$$

are both valid, as demonstrated by

$$\dfrac{\dfrac{A \vee B \text{ false} \quad \dfrac{\dfrac{A : \text{prop}}{A \text{ true } \ (A \text{ true})}}{A \vee B \text{ true } \ (A \text{ true})}}{\Lambda \text{ true } \ (A \text{ true})}}{A \text{ false}} \, ,$$

[23]Though not explicitly stated in this form, this insight is due to Brouwer, 'The Unreliability of the Logical Principles', p. 110.

and similarly with A and B interchanged. The double negative form of the law of excluded middle is now demonstrated by

$$\cfrac{\cfrac{\cfrac{A : \mathrm{prop}}{A \vee {\sim}A \text{ false } \ (A \vee {\sim}A \text{ false})}}{{\sim}A \text{ false } \ (A \vee {\sim}A \text{ false})} \quad \cfrac{\cfrac{A : \mathrm{prop}}{A \vee {\sim}A \text{ false } \ (A \vee {\sim}A \text{ false})}}{A \text{ false } \ (A \vee {\sim}A \text{ false})}}{\cfrac{\Lambda \text{ true } \ (A \vee {\sim}A \text{ false})}{A \vee {\sim}A \text{ irrefutable}}} .$$

If we combine the irrefutability of $A \vee {\sim}A$ with proof by dilemma we get the inference rule

$$\frac{B \text{ irrefutable } \ (A \text{ true}) \quad B \text{ irrefutable } \ (A \text{ false})}{B \text{ irrefutable}} ,$$

which may be termed proof by cases. Note that we may use the strong assumption that A is true in the leftmost proof that B is irrefutable, just as we did for implication introduction.

This shows that the laws of propositional logic are valid also when reasoning about irrefutability instead of truth, and that proof by contradiction and proof by cases may be used when proving an irrefutable conclusion.

§ 2. The law of excluded middle

Logicians make a distinction between the law of excluded middle and the principle of bivalence. The law of excluded middle is usually formulated as the proposition $A \vee {\sim}A$ being true whenever A is a proposition. It is natural to equate this law with the (invalid) inference rule

$$\frac{A : \mathrm{prop}}{A \vee {\sim}A \text{ true}}$$

in intuitionistic type theory. The principle of bivalence cannot be formulated as an inference rule in intuitionistic type theory—it has to be formulated in the metalanguage: for any proposition A, either A is true *or* ${\sim}A$ is true, with a metalinguistic *or*.

The validity of the principle of bivalence of course depends on the meaning assigned to the notions of proposition, truth, and negation, and the exact sense in which exactly one of A and ${\sim}A$ must be true; the validity of the law of excluded middle further depends on the meaning assigned to disjunction. Under the bivalent truth value interpretation of the notions involved,[24] both principles are valid.

For the remainder of this section, let proposition, truth, negation, and disjunction have their intuitionistic meaning, as is given above, i.e., a proposition is identified with a set, a true proposition with an inhabited set, the negation of A is defined as $A \supset \Lambda$, and a cause of $A \vee B$ consists of a cause of A or a cause of B, together with information about which cause it is that is given. To

[24] According to which a proposition is interpreted as a truth value (i.e., as an element of the set B); the truth of a proposition A is interpreted as A being equal to *the true* (i.e., as A being equal to 1); and negation and disjunction have their usual Boolean definitions (truth tables).

claim that the law of excluded middle is valid, we need to know a cause of the proposition $A \vee \sim A$ for any proposition A. Such a cause is in effect a method for deciding whether A or $\sim A$ is true for any proposition A. It is important that, to claim that the law of excluded middle is valid, one need to *actually know*, or possess, a method for deciding any proposition, not merely believe that such method exists, in whatever sense of the word. Since I do not possess any such method, and am confident that it is not possessed by anybody else either, I call the law of excluded middle invalid.

A proposition A is *decidable*, abbreviated 'A decidable', if the proposition $A \vee \sim A$ is true, and *stable* (under double negation), abbreviated 'A stable', if the proposition $\sim\sim A \supset A$ is true. These definitions are expressed by the bidirectional inference rules

$$\frac{A \vee \sim A \text{ true}}{A \text{ decidable}} \qquad \text{and} \qquad \frac{\sim\sim A \supset A \text{ true}}{A \text{ stable}} .$$

The law of excluded middle is tantamount to the principle that any proposition is decidable. The relation between *decidable* and *stable* is that any decidable proposition is stable, i.e.,

$$\frac{A \text{ decidable}}{A \text{ stable}} , \tag{4}$$

and that if $A \vee \sim A$ is stable, then A is decidable, i.e.

$$\frac{A \vee \sim A \text{ stable}}{A \text{ decidable}} . \tag{5}$$

The second of these inference rules follows directly by *modus ponendo ponens*, since $A \vee \sim A$ is irrefutable, i.e.,

$$\frac{A \vee \sim A \text{ stable} \quad A \vee \sim A \text{ irrefutable}}{A \text{ decidable}} .$$

In general, we have

$$\frac{A \text{ stable} \quad A \text{ irrefutable}}{A \text{ true}} ,$$

i.e., that A is stable is precisely what is needed to infer the truth of A from the irrefutability of A. Inference rule (4) is demonstrated by

$$\frac{A \text{ decidable} \quad \dfrac{\dfrac{A : \text{prop}}{A \text{ true} \ \ (A \text{ true}, A \text{ irrefutable})}}{A \text{ stable} \ \ (A \text{ true})} \quad \dfrac{\dfrac{\dfrac{A : \text{prop}}{\Lambda \text{ true} \ \ (A \text{ false}, A \text{ irref.})}}{A \text{ true} \ \ (A \text{ false}, A \text{ irref.})}}{A \text{ stable} \ \ (A \text{ false})}}{A \text{ stable}} .$$

From inference rule (5), it also follows that if any proposition is stable, then any proposition is decidable; since if *any* proposition is stable, then, *a fortiori*, any proposition of the form $A \vee \sim A$ is stable, so that A, which is arbitrary, is decidable.

Having made these distinctions, it can be explained how nearly all logicians from Aristotle to Brouwer could consider the law of excluded middle as self-evident and tantamount to the principle of noncontradiction. To my mind, it boils down to a confusion between the different senses of the word *or*.[25] If proof by dilemma is to be an inference rule productive of scientific knowledge, i.e., if its conclusion is to be *true* in our sense, then disjunction has to have its intuitionistic meaning. The sense of disjunction that makes the law of excluded middle, as it were, equivalent to the principle of noncontradiction is defined by

$$A \vee_w B \stackrel{\text{def}}{=} \sim(\sim A \mathbin{\&} \sim B) : \text{prop}.^{26}$$

Clearly, the inference rule

$$\frac{A : \text{prop}}{A \vee_w \sim A \text{ true}}$$

is valid. Note also that the proposition $A \vee_w \sim A$ is stable since it is negative. With this definition, disjunction introduction is valid with a merely irrefutable premiss, i.e., the inference rules

$$\frac{A \text{ irrefutable} \quad B : \text{prop}}{A \vee_w B \text{ true}} \quad \text{and} \quad \frac{A : \text{prop} \quad B \text{ irrefutable}}{A \vee_w B \text{ true}}$$

are valid, as demonstrated by

$$\cfrac{\cfrac{A \text{ irrefutable} \qquad \cfrac{\cfrac{A : \text{prop} \quad B : \text{prop}}{\sim A \mathbin{\&} \sim B \text{ true} \ \ (\sim A \mathbin{\&} \sim B \text{ true})}}{A \text{ false} \ \ (\sim A \mathbin{\&} \sim B \text{ true})}}{\Lambda \text{ true} \ \ (\sim A \mathbin{\&} \sim B \text{ true})}}{A \vee_w B \text{ true}},$$

and similarly with B and A interchanged. Moreover, proof by dilemma, i.e.,

$$\frac{A \vee_w B \text{ true} \quad C \text{ irrefutable} \ \ (A \text{ true}) \quad C \text{ irrefutable} \ \ (B \text{ true})}{C \text{ irrefutable}},$$

is valid for this definition of disjunction, but produces only an irrefutable conclusion. This inference rule is demonstrated by

$$\cfrac{\cfrac{A \vee_w B \text{ true} \qquad \cfrac{\cfrac{\cfrac{C \text{ irrefutable} \ \ (A \text{ true})}{\Lambda \text{ true} \ \ (C \text{ false}, A \text{ true})}}{A \text{ false} \ \ (C \text{ false})} \quad \cfrac{\cfrac{C \text{ irrefutable} \ \ (B \text{ true})}{\Lambda \text{ true} \ \ (C \text{ false}, B \text{ true})}}{B \text{ false} \ \ (C \text{ false})}}{\sim A \mathbin{\&} \sim B \text{ true} \ \ (C \text{ false})}}{\Lambda \text{ true} \ \ (C \text{ false})}}{C \text{ irrefutable}}.$$

Similarly, *modus tollendo ponens* only produces knowledge of an irrefutable proposition with this definition of disjunction.

There is also a third possible definition of disjunction which arises from taking the rule of *modus tollendo ponens* as the principal way of using a disjunctive major premiss, instead of proof by dilemma, as is done for intuitionistic

[25] In fact, there are two confusions involved: the first confusion is between *inclusive or* and *exclusive or*; the latter excludes the case of both disjuncts being true and could be defined by $(A \vee B) \mathbin{\&} \sim(A \mathbin{\&} B)$. In what follows I address only the different senses of inclusive or.

[26] I will use the symbol '$\vee_w$' for *weak disjunction* and the symbol '$\vee_m$' for *middle disjunction*.

disjunction. To simplify matters, I simply define this notion of disjunction in terms of implication, by

$$A \vee_m B \overset{\text{def}}{=} ({\sim}A \supset B) \,\&\, ({\sim}B \supset A) : \text{prop},$$

from which it is clear that *modus tollendo ponens* is valid with the proposition $A \vee_m B$ as major premiss and with a true conclusion. Disjunction introduction is valid also with this definition of disjunction, i.e., the inference rules

$$\dfrac{A\ \text{true} \quad B : \text{prop}}{A \vee_m B\ \text{true}} \quad \text{and} \quad \dfrac{A : \text{prop} \quad B\ \text{true}}{A \vee_m B\ \text{true}}$$

are valid, as demonstrated by

$$\dfrac{\dfrac{\dfrac{\dfrac{A\ \text{true} \quad \dfrac{A : \text{prop}}{A\ \text{false}\ (A\ \text{false})}}{\Lambda\ \text{true}\ (A\ \text{false})}}{B\ \text{true}\ (A\ \text{false})}}{{\sim}A \supset B\ \text{true}} \qquad \dfrac{\dfrac{A\ \text{true} \quad B : \text{prop}}{A\ \text{true}\ (B\ \text{false})}}{{\sim}B \supset A\ \text{true}}}{A \vee_m B\ \text{true}},$$

and similarly with A and B interchanged. Moreover, by definition, we have

$$A \vee_m {\sim}A = ({\sim}A \supset {\sim}A) \,\&\, ({\sim}{\sim}A \supset A) : \text{prop},$$

which, since the first conjunct is trivially true, is equivalent to ${\sim}{\sim}A \supset A$, i.e., to A being stable under double negation. So, with this interpretation of disjunction, the law of excluded middle becomes tantamount to the principle that any proposition is stable under double negation. As shown above, this is tantamount to the principle that any proposition is decidable, i.e., to the strong or intuitionistic law of excluded middle, which is invalid.

The three senses of disjunction $\vee$, $\vee_w$, and $\vee_m$, will be called, respectively, *strong*, *weak*, and *middle* disjunction. Some people claim that as the intuitionists deny the law of excluded middle, which they do in its strong form, they must affirm a third truth value between truth and falsity, a consequence which would follow from a denial of the weak law of excluded middle. The law of excluded middle is sometimes called *tertium non datur*, literally meaning that a third is not given, i.e., a third option between truth and falsity; now, that a third *is* given means that the proposition $A \vee {\sim}A$ is false, whence that a third *is not* given means that $A \vee {\sim}A$ is irrefutable. With this interpretation, the law of excluded middle is *not* equivalent to *tertium non datur*, as the latter is intuitionistically valid but the former is not.

As for the third truth value, it might well be that the antagonists of intuitionism are referring to the state of doubt. With respect to knowledge, a man's attitude towards a proposition can be broadly divided into three: he may know the proposition to be true, he may know the proposition to be false, and he may know neither that it is true nor that it is false. Thus, true, doubtful, and false, are not three truth values, but three knowledge states, as it were.

It remains to investigate the principle of bivalence under the intuitionistic interpretation of the notions involved. In the very beginning of *Outlines of Pyrrhonism*, Sextus Empiricus makes the following observation:

"The natural result of any investigation is that the investigators either discover the object of search or deny that that it is discoverable and confess it to be inapprehensible or persist in their search. So, too, with regard to the objects investigated by philosophy, this is probably why some have claimed to have discovered the truth, others have asserted that it cannot be apprehended, while others again go on inquiring."[27]

Sextus Empiricus calls these three views dogmatic, academic, and sceptic, respectively—Sextus Empiricus himself of course being a sceptic. The three possible outcomes of the search for an object are, in particular, applicable to the search for a cause of the truth of a proposition, and correspond to the three knowledge states mentioned above. Further analysis of this argument reveals more, viz., that, for scientific knowledge, the only possible changes in attitude are from doubtful to true and from doubtful to false. So doubt has a special status among the knowledge states in that it is possible to overcome.

I will take the principle of bivalence to be tantamount to the principle that all doubt is possible to overcome: *non ignorabimus* to speak with Hilbert.[28] Here are some possible attitudes towards this principle.

(1) The most optimistic position is that there is a systematic method to establish either A true or $\sim A$ true, for any proposition A.[29] I take this position to imply a positive solution to Hilbert's Entscheidungsproblem, in direct contradiction with the result gained by Church and Turing.[30] Thus, this position is self-contradictory.

(2) The second most optimistic position is to claim to know a method to establish either A true or $\sim A$ true, for any proposition A. Somebody in this position claims to have evidence for the strong law of excluded middle. This certainly entails the principle of bivalence since, if the intuitionistic disjunction $A \vee B$ is true, then A is true or B is true. I will call anybody in possession of a method for deciding any proposition an oracle.[31] There seems to be no systematic way of refuting somebody who claims to be an oracle, but there are ample reasons to be sceptical of such a claim.

(3) A third possibility is to claim that there is a method to establish either A true or $\sim A$ true, for any proposition A, without claiming to know such a method, i.e., to claim that there are oracles, without claiming to be one.

(4) A fourth position is that there *may* be a method to establish either A true or $\sim A$ true, for any proposition A, but that this method is not humanly attainable, i.e., the content of this position is that there are no human oracles.

(5) A fifth and less optimistic position is that there is no method which, for any proposition A, establishes either A true or $\sim A$ true, i.e., that there cannot

[27]Empiricus, *Outlines of Pyrrhonism*, Bk. 1, Ch. 1.

[28]Cf. Hilbert, 'Mathematical problems', p. 445.

[29]In this list, the word method is to be understood in the sense defined in Ch. IV, and the systematic method as a definite method, or algorithm. Note also that the schematic letter A stands for a *closed* proposition.

[30]Cf. Church, 'An unsolvable problem of elementary number theory' and Turing, 'On Computable Numbers'.

[31]The use of the word *oracle* in this connection was introduced by Turing, 'Systems of logic based on ordinals', § 4, p. 172.

be any oracles.

(6) Finally, the least optimistic position is that there is a proposition A for which it can be known to be impossible to establish A true and equally impossible to establish $\sim A$ true. This position is self-contradictory if we agree that we may infer that $\sim A$ is true from knowledge of the impossibility of establishing that A is true.[32] This entailment is reasonable since to know that it is impossible to establish A, one has to possess a method of producing an absurd consequence from an alleged cause of A, and this method is a cause of $\sim A$. So, for the alleged counterexample A to the principle of bivalence, we have A false and A irrefutable, which is absurd.

The above six positions all say *something* on the principle of bivalence. Yet another option is to have no opinion about it—perhaps this is the most prudent position, since nothing can be said with certainty either for or against positions two, three, four, and five.

This talk about oracles in connection with the principle of bivalence brings us to a related topic where this principle has been discussed, namely περὶ δυνατῶν, about things possible.[33] My sources for this discussion are Aristotle's *Perihermenias* and Cicero's *De Fato*. To establish the connection between the principle of bivalence and fate, it suffices to apply the six positions on the principle of bivalence, discussed above, to propositions about the future. It might be objected that in the previous discussion it was implicit that the propositions were non-temporal, but, as the connection with future propositions is interesting in itself, I will pursue it nevertheless. According to Cicero, the ancients argued that if something was without cause, this would contradict the principle that every proposition was necessarily either true or false.[34] In our terminology, that something, A, is without cause can be interpreted as A being irrefutable without having a cause; this cannot happen if the principle of bivalence holds, because then $\sim A$ must have a cause if A does not, in contradiction to the assumption that A was irrefutable. Thus, the principle of bivalence implies that no fact, i.e., irrefutable proposition, is without cause. Cicero reports that, from this implication, Chrysippus argued, by *modus ponendo ponens*, that all things take place by fate, and Epicurus, by *modus tollendo tollens*, that not every proposition is necessarily either true or false:

"At this point, in the first place if I chose to agree with Epicurus and to say that not every proposition is either true or false, I would rather suffer that nasty knock than agree that all events are caused by fate; for the former opinion has something to be said for it, but the latter is intolerable."[35]

Now, that Chrysippus' position is intolerable shows that to avoid fatalism, we have to deny the principle of bivalence, i.e., we have to take position five above, at least if we take the notion of proposition in the most general possible sense, including propositions about the future, and take the principle

---

[32]Cf. Martin-Löf, 'Verificationism Then and Now', Third Law, p. 16.
[33]Cicero, *De Fato*, Ch. 1.
[34]Ibid., Ch. 10, beginning.
[35]Ibid., Ch. 10, n. 21.

of bivalence to mean that either A is true *now* or $\sim A$ is true *now*.[36] This of course does not settle the question whether the principle of bivalence holds for that which is actual or for that which is timeless, like mathematics. Aristotle escapes the problem by making this distinction:

"For one half of the said contradiction must be true and other half false. But we cannot say which half is which. Though it may be that one is more probable, it cannot be true yet or false. There is evidently, then, no necessity that one should be true, the other false, in the case of affirmations and denials. For the case of those things which as yet are potential, not actually existent, is different from that of things actual."[37]

This can be read as a denial of the most general form of the principle of bivalence, while maintaining that it holds for propositions about the present, i.e., about things actual.

To maintain the principle of bivalence for actual propositions, i.e., that every proposition about the present is either true or false, entails that every proposition about the future *will become* either true or false. If we accept this principle we have to beware of an error which is easy to make, viz., to claim that if two persons hold contradictory propositions about the future, one of them is right and the other wrong. It is not so, because to know is to know by causes, and, most likely, both of them are wrong, i.e., speaking without knowing. Put differently, if you make a guess, and it turns out as you predicted, your guess was still not knowledge, i.e., you did not speak the truth. This kind of reasoning seems to have confused Cicero:

"For it is necessary that of two contradictory propositions, *pace* Epicurus, that one should be true and the other false; for example, 'Philoctetes will be wounded' was true, and 'Philoctetes will not be wounded' false, for the whole of the ages of the past; unless perhaps we choose to follow the opinion of the Epicureans, who say that propositions of this sort are neither true nor false, or else, when ashamed of that, they nevertheless make the still more impudent assertion that disjunctions consisting of contradictory propositions are true, but that the statements contained in the propositions are neither of them true. What marvellous effrontery and pitiable ignorance of logical method! For if anything propounded is neither true nor false, it certainly is not true; but how can something that is not true not be false, or how can something that is not false not be true?"[38]

It is interesting to note that the position of the intuitionists agrees rather well with that of the Epicureans, as reported by Cicero: they deny the strong law of excluded middle, i.e., the truth of the proposition $A \vee \sim A$, and, when ashamed of that, affirm the weak law of excluded middle, i.e., the truth of the proposition $A \vee_w \sim A$, and deny the principle of bivalence. The final sentence of Cicero's argument can be seen as a refutation of position six, above,

[36]This is the most natural interpretation of the principle of bivalence, since the assertion A true can be expanded into *I know a logical cause of A*, in which the *now* is implicit.
[37]Aristotle, *Perih.*, Ch. 9, 19a37–19b5.
[38]Cicero, *De Fato*, Ch. 16, nn. 37–38. I have changed the translation to conform with standard terminology in logic by replacing the word contrary with the word contradictory and removing Cicero's comment giving an explanation of his unusual sense of the word contrary.

showing that one has to be careful when formulating a denial of the principle of bivalence.

A final objection to the principle of bivalence and the law of excluded middle, this time even for propositions about the present and the timeless, is that it fails to hold because of an intrinsic vagueness in the terms involved in the proposition at hand.[39] Problems of this kind are related to the old paradoxes about the bald man and the heap:[40] How many hairs may a man have and still be called bald? How many stones make a heap? If, for every number n, the proposition *n stones make a heap* is either true or false, there must be a least number for which it is true, contrary to intuition. To get the unintuitive conclusion, we have to use the law of excluded middle. An often overlooked virtue of intuitionism is that it dissolves this kind of paradoxes: I can affirm that one or two stones do not make a heap and that fifty or more stones make a heap without having to make up my mind for the numbers in between.[41]

§ 3. The philosophy of mathematics

Today, mathematicians generally do not pay much attention to philosophical issues. The reason for this can be traced back to the largely unresolved issues concerning the nature of mathematical entities which were fiercely debated around the turn of the last century.[42] I think that the most significant outcome of this debate was that intuitionism was rejected by all but a few. Already in 1927, Weyl made the following observation:

"If Hilbert's view prevails over intuitionism, as appears to be the case, *then I see in this a decisive defeat of the philosophical attitude of pure phenomenology*".[43]

Indeed, Hilbert's view did prevail, and in 1970 the authors behind the pseudonym Bourbaki could write:

"The intuitionist school, whose memory will undoubtedly survive only as a historical curiosity, has at least rendered the service of having obliged its opponents, that is to say the vast majority of mathematicians, to clarify their own positions and to become more consciously aware of the reasons (whether logical or sentimental) for their confidence in mathematics."[44]

I think that intuitionism has to be reevaluated as a tenable approach to the foundation of mathematics, now that Bishop has purged it from the points on which it was in conflict with classical mathematics.[45] I have said enough in defense of intuitionism (or constructivism—I take the two to be synonymous)

---

[39]Cf. Geach, 'The law of excluded middle', pp. 71-73.

[40]Kneale and Kneale, *The Development of Logic*, p. 114.

[41]To be more precise: one could define the predicate heap(n) inductively by stipulating that h(m) : el(heap($m + 50$)), but without admitting the strongest possible elimination rule for this predicate: so that heap(n) can be demonstrated for $n > 49$, using h, and that ∼heap(n) can be demonstrated for $n < 3$, using a weak form of the elimination rule.

[42]We could delimit this period in time, the Grundlagenkrise, as starting in 1879 with the publication of Frege's *Begriffsschrift*, and ending in 1931 with Gödel's publication of his incompleteness theorems.

[43]Weyl, 'Comments on Hilbert's second lecture on the foundations of mathematics', p. 484.

[44]Bourbaki, *Elements of Mathematics*, p. 336.

[45]Cf. Bishop and Bridges, *Constructive Analysis*, Ch. 1.

above, so it remains to explain why the competing approaches to the foundations of mathematics, namely, formalism and set-theoretical Platonism, are untenable. The following analysis is given by Simpson:

"We have mentioned three competing 20th century doctrines: formalism, constructivism, set-theoretical Platonism. None of these doctrines are philosophically satisfactory, and they do not provide much guidance for mathematically oriented scientists and other users of mathematics. As a result, late 20th century mathematicians have developed a split view, a kind of Kantian schizophrenia, which is usually described as "Platonism on weekdays, formalism on weekends". In other words, they accept the existence of infinite sets as a working hypothesis in their mathematical research, but when it comes to philosophical speculation, they retreat to a formalist stance. Thus they have given up hope of an integrated view which accounts for both mathematical knowledge and the applicability of mathematics to physical reality. In this respect, the philosophy of mathematics is in a sorry state."[46]

First I consider set-theoretical Platonism. The following quotation from Bishop is illuminating:

"The fact that space has been arithmetized loses much of its significance if space, number, and everything else are fitted into a matrix of idealism where even the positive integers have an ambiguous computational existence. Mathematics becomes the game of sets, which is a fine game as far as it goes, with rules that are admirably precise. The game becomes its own justification, and the fact that it represents a highly idealized version of mathematical existence is universally ignored."[47]

Skolem's explanation of why he had not previously published his, still relevant, critique of extensional set theory is also of interest:

"I believed that it was so clear that axiomatization in terms of sets was not a satisfactory ultimate foundation of mathematics that mathematicians would, for the most part, not be very much concerned with it. But in recent times I have seen to my surprise that so many mathematicians think that these axioms of set theory provide the ideal foundation for mathematics; therefore it seemed to me that the time had come to publish a critique."[48]

To be more exact, let me identify the following four problems with extensional set theory in general, and Zermelo-Fraenkel set theory in particular.

(1) Lack of evidence for consistency.

(2) Lack of constructivity.

(3) Lack of agreement with other sciences.

(4) Lack of typing.

Lack of evidence for consistency. One could say that the designers of extensional set theory (Zermelo, et al.) opted for the strongest system not evidently inconsistent. The more precarious approach is to look for the strongest system

[46]Simpson, 'Logic and mathematics', § 3.2.
[47]Bishop and Bridges, *Constructive Analysis*, Ch. 1, p. 7.
[48]Skolem, 'Some remarks on axiomatized set theory', pp. 300–301.

which is evidently consistent, as is done in intuitionistic type theory. The latter system turns out to be much weaker than the former.

Lack of constructivity. This point has to do with the nature of mathematical existence. In a nonconstructive setting, a proof of the existence of a mathematical object with a certain property does not entail that such an object can be constructed. That is, there is no general method pass from knowledge of the truth of the proposition $\sim\sim(\exists x : A)P(x)$ to knowledge of the truth of the proposition $(\exists x : A)P(x)$.

Two historically particularly important examples of nonconstructive proofs are the proof of the existence of transcendental numbers by means of a cardinality argument,[49] and Hilbert's proof of his basis theorem.[50] That such proofs were debated from the very day they appeared shows, if not that they are invalid, then at least that there is something intrinsically unsatisfactory about them. Whether one accepts them or not, it has to be admitted that they destroy the classical notion of function. Computationally unfounded appeal to the law of excluded middle, such as in a nonconstructive existence proof, corrupts the notion of function, as explained in Chapter IV.

Lack of agreement with other sciences. The set theoretical principle that everything is a set leads to discrepancies between the meanings of words as used in mathematics and the meanings of the same words as used in other sciences. From ancient times to the time of Dedekind,[51] the concept of real number, continuous quantity, or magnitude, was the same in mathematics and the rest of science, as well as in everyday discourse. The difference in understanding was only one of degree.

Lack of typing. The last argument is directed against the view that everything is a set. The meaningfulness of taking the intersection of, say, the real number π, and the transcendental function sin, is questionable, to say the least.[52] Computer programmers know that types are indispensable for correctness.

Formalism, on the other hand, fails to account for the contentfulness of mathematics since, according to the formalist, mathematics does not have any definite subject matter.

"This is the predicament of formal arithmetic: it cannot help but make use of sentences supposed to express thoughts, but nobody can determine exactly what these thoughts are."[53]

Most people would agree that at least classical mathematics has a definite subject matter, and that the mathematical symbolism can be given a nonstandard interpretation does not alter this.

---

[49]This proof is commonly attributed to Cantor, but first published by Klein (cf. Gray, 'Georg Cantor and Transcendental Numbers').
[50]Hilbert, 'Über die Theorie der algebraischen Formen'.
[51]Dedekind, 'Continuity and Irrational Numbers'.
[52]For a detailed account of this argument, see de Bruijn, 'On the roles of types in mathematics'.
[53]Frege, *Grundgesetze der Arithmetik II*, § 105 (trans. Black).

However, there are also many things about which formalists and intuitionists agree. Consider Curry's three criteria for the acceptability of a formal system:

(1) The intuitive evidence of the premises;

(2) consistency (an internal criterion);

(3) the usefulness of the theory as a whole.[54]

To my mind, (2) is a criterion *sine qua non* for the acceptability of a formal system. This view is not incompatible with formalism (it was embraced by Hilbert) even though Curry had a more "evolutionary" point of view. With regard to points (1) and (3), it seems as if the difference between intuitionists and formalists is one of degree: the intuitionists emphasizing (1), sometimes at the expense of (3), and the formalists the other way around. Clearly, some sort of balance must be sought between them, when they are in conflict, as both are goals worth striving for, but (1) should not be given up easily. By this I simply mean that if nothing can be said with complete certainty, and something has to be said, then (1) has to give way.

The foundations of mathematics must account both for mathematical knowledge, i.e. for mathematics as classically understood, and for the applicability of mathematics, i.e. for our ability to use mathematics to reason about the world. Certainly, the success of mathematical methods in the 20th century shows that modern mathematics is applicable, but its philosophical foundation fails to explain why.

I hope that the increasing impact of computer science on mathematics will make mathematicians realize that the alleged paradise, created by Cantor, is not a real paradise after all,[55] so that mathematics, once again, can be held up as the ideal of certainty and clarity; only then can mathematics form the core of a *lingua characteristica* with any hope of embracing the less exact sciences.

Qui potest capere capiat!

[54]Taken from Curry, 'Remarks on the definition and nature of mathematics', p. 232.

[55]Alluding to what Hilbert wrote in 1926: "Aus dem Paradies, das Cantor uns geschaffen, soll uns niemand vertreiben können" ('Über das Unendliche', p. 170): "No one will drive us from the paradise which Cantor created for us" (trans. Bauer-Mengelberg).

Bibliography

Abadi, M. et al. 'Explicit substitution'. In: *J. Funct. Program.* 1.4 (1991), pp. 375–416.

Abelson, H. and G. J. Sussman. *Structure and Interpretation of Computer Programs.* 2nd ed. MIT Press and McGraw-Hill, 1996.

Ackermann, W. 'Zum Hilbertschen Aufbau der reellen Zahlen'. In: *Math. Ann.* 99.1 (1928), pp. 118–133.

Aczel, P. 'The type theoretic interpretation of constructive set theory'. In: *Logic Colloquium '77.* Ed. by A. Macintyre, L. Pacholski and J. Paris. Amsterdam: North-Holland, 1978.

Aquinas, St. T. 'Summa Theologiae'. In: *Opera omnia. Iussu impensaque Leonis XIII P. M. edita.* Vols. 4–12. Rome: Ex typographia polyglotta S. congregationis de propaganda fide, 1888–1906.

— *In duodecim libros Metaphysicorum Aristotelis expositio.* Ed. by M.-R. Cathala and R. M. Spiazzi. Turin: Marietti, 1950.

— 'De Veritate'. In: *Quaestiones Disputatae.* Ed. by R. Spiazzi. Vol. 1. Turin-Rome: Marietti, 1953.

— 'In Peri Hermeneias'. In: *In Aristotelis Libros Peri Hermeneias et Posteriorum Analyticorum Expositio.* 2nd ed. Marietti, 1964.

— 'In Posteriorum Analyticorum'. In: *In Aristotelis Libros Peri Hermeneias et Posteriorum Analyticorum Expositio.* 2nd ed. Marietti, 1964.

— 'De Potentia'. In: *Quaestiones Disputatae.* Ed. by P. M. Pession. 10th ed. Vol. 2. Turin-Rome: Marietti, 1965, pp. 1–276.

— 'De ente et essentia'. In: *Opera omnia. Iussu impensaque Leonis XIII P. M. edita.* Vol. 43. Rome: Editori di San Tommaso, 1976, pp. 315–381.

Aristotle. *Politics.* Trans. by H. Rackham. Loeb Classical Library 264. Harvard University Press, 1932.

— *Metaphysics.* Trans. by H. Tredennick. Loeb Classical Library 271 & 287. Harvard University Press, 1933 & 1935.

— *Analytica priora.* Trans. by H. Tredennick. Loeb classical library 325. Harvard University Press, 1938, pp. 181–531.

— *Categories.* Trans. by H. P. Cooke. Loeb classical library 325. Harvard University Press, 1938, pp. 9–109.

— *Perihermenias.* Trans. by H. P. Cooke. Loeb classical library 325. Harvard University Press, 1938, pp. 111–179.

— *Analytica posteriora.* Trans. by H. Tredennick. Loeb classical library 391. Harvard University Press, 1960, pp. 1–261.

— *Topica.* Trans. by E. S. Forster. Loeb Classical Library 391. Harvard University Press, 1960, pp. 263–739.

J.G. Granström, *Treatise on Intuitionistic Type Theory*, Logic, Epistemology, and the Unity of Science 22, DOI 10.1007/978-94-007-1736-7,

Aristotle. *The Physics*. Trans. by P. H. Wicksteed and F. M. Cornford. Loeb Classical Library 228 & 255. Harvard University Press, 1986 & 2006.

Arnauld, A. *The art of thinking*. Trans. by J. Dickoff and P. James. New York: Bobbs-Merrill, 1964.

Augustine, A. *De Dialectica*. Ed. by J. Pinborg. Trans. by B. D. Jackson. Synthese Historical Library. Springer, 1975.

Austin, J. L. *How to Do Things with Words*. Ed. by J. O. Urmson and M. Sbisà. 2nd ed. Harvard University Press, 1975.

Backus, J. W. 'Can programming be liberated from the von Neumann style? A functional style and its algebra of programms'. In: *Communications of the ACM* 21.8 (1978), pp. 613–641.

— 'Function Level Programs as Mathematical Objects. Proceedings of the 1981 conference on Functional programming languages and computer architecture'. In: *FPCA '81*. New York: ACM, 1981, pp. 1–10.

Bacon, F. *Novum Organum*. London: J. Billius, 1620.

Becker, O. 'Mathematische Existenz. Zur Logik und Ontologie mathematischer Phänomene'. In: *Jahrbuch für Philosophie und phänomenologische Forschung*. Vol. 8. 1927, pp. 441–809.

Bellantoni, S. and S. Cook. 'A new recursion-theoretic characterization of the polytime functions'. In: *Comput. Complex.* 2.2 (1992), pp. 97–110.

Bernays, P. 'Die Philosophie der Mathematik und die Hilbertsche Beweistheorie'. In: *Blätter für Deutsche Philosophie* 4 (1931), pp. 326–367.

— 'Mathematische Existenz und Widerspruchsfreiheit'. In: *Etudes de Philosophie des sciences en hommage à Ferdinand Gonseth*. Neuchatel: Griffon, 1950, pp. 11–25.

Bishop, E. and D. Bridges. *Constructive Analysis*. Springer, 1985.

Blancanus, J. 'A Treatise on the Nature of Mathematics along with a Chronology of Outstanding Mathematicians'. In: Mancosu, P. *Philosophy of Mathematics and Mathematical Practice in the Seventeenth Century*. Trans. by G. Klima. Oxford University Press, 1996, pp. 178–212.

Bocheński, I. M. *Ancient Formal Logic*. Amsterdam: North-Holland, 1951.

Boëthius, S. 'De syllogismo categorico'. In: *Patrologiae cursus completus. Series Latina*. Ed. by J.-P. Migne. Vol. 64. 793C–832A. Paris: Garnier, 1847.

— 'De hypotheticis syllogismis'. In: *Logicalia*. Ed. by L. Obertello. Vol. 1. Parma: Brescia, 1968.

Bolzano, B. *Wissenschaftslehre*. Sulzbach: Seidel, 1837.

— *Paradoxes of the Infinite*. Trans. by D. A. Steele. London: Routledge, 1950.

Boole, G. 'The Calculus of Logic'. In: *Cambridge and Dublin Math. J.* 3 (1848), pp. 183–198.

— *An Investigation of the Laws of Thought*. London: Walton and Maberly, 1854.

Bourbaki, N. *Elements of Mathematics. Theory of Sets*. Addison-Wesley, 1968.

Brouwer, L. E. J. 'Intuitionism and formalism'. Trans. by A. Dresden. In: *Bull. Amer. Math. Soc.* 20.2 (1913), pp. 81–96.

— 'Intuitionistische Zerlegung mathematischer Grundbegriffe'. In: *Jahresber. d. Deutsch. Math.-Vereinig.* 33 (1925), pp. 251–256.

— 'The Unreliability of the Logical Principles'. In: *Collected Works. Philosophy and Foundations of Mathematics.* Ed. by A. Heyting. Vol. 1. Amsterdam: North Holland, 1975 (original from 1908), pp. 107–111.
Brown, H. C. 'The Logic of Mr. Russell'. In: *J. Philos. Psychol. Sci. Meth.* 8.4 (1911), pp. 85–91.
de Bruijn, N. G. *Automath, a language for mathematics.* Tech. rep. 68-WSK-05. Technological University Eindhoven, 1968.
— 'A survey of the project Automath'. In: *To H. B. Curry: Essays in Combinatory Logic, Lambda Calculus and Formalism.* Academic Press, 1980, pp. 589–606.
— *Recommendations concerning standardization of mathematical formulas.* Tech. rep. M11. Technological University Eindhoven, 1980.
— 'Telescopic mappings in typed lambda calculus'. In: *Inform. Comput.* 91 (1991), pp. 189–204.
— 'On the roles of types in mathematics'. In: *The Curry-Howard isomorphism.* Ed. by P. de Groote. Academia, 1995, pp. 27–54.
Buridan, J. *Summulae de Dialectica.* Trans. by G. Klima. Yale University Press, 2001.
de Bury, R. *Philobiblon.* Ed. by A. Nelson. Stockholm: Föreningen för Bokhandtverk, 1922.
Cajetan, T. *Commentaria in Praedicamenta Aristotelis.* Rome: Angelicum, 1939.
Cajori, F. *A History of Mathematics.* New York: Macmillan, 1919.
— 'Mathematical Signs of Equality'. In: *Isis* 5.1 (1923), pp. 116–125.
— *A History of Mathematical Notations.* Vol. 1. Illinois: Open Court, 1928.
— *A History of Mathematical Notations.* Vol. 2. Illinois: Open Court, 1929.
Cantor, G. 'Über unendliche, lineare Punktmannichfaltigkeiten'. In: *Math. Ann.* 20.1 (1882), pp. 113–121.
— 'Beiträge zur Begründung der transfiniten Mengenlehre'. In: *Math. Ann.* 46.4 (1895), pp. 481–512.
Cardelli, L. and P. Wegner. 'On Understanding Types, Data Abstraction, and Polymorphism'. In: *ACM Computing Surveys* 17.4 (1985), pp. 471–522.
Carnap, R. *Introduction to Symbolic Logic and Its Applications.* New York: Dover, 1958.
Carroll, L. 'What the Tortoise said to Achilles'. In: *Mind* 4.14 (1895), pp. 278–280.
— *Symbolic Logic.* London-New York: Macmillan & Co., 1896.
Cartmell, J. 'Generalized algebraic theories and contextual categories'. In: *Ann. of Pure Appl. Logic* 32 (1986), pp. 209–243.
Church, A. 'On the law of excluded middle'. In: *Bull. Amer. Math. Soc.* 34 (1928), pp. 75–78.
— 'A bibliography of symbolic logic'. In: *J. Symb. Log.* 1 (1936), pp. 121–218.
— 'An unsolvable problem of elementary number theory'. In: *Amer. J. Math.* 58.2 (1936), pp. 345–363.
— 'A Formulation of the Simple Theory of Types'. In: *J. Symb. Log.* 5 (1940), pp. 56–68.
Cicero, M. T. *De Fato.* Trans. by H. Rackham. Loeb Classical Library. Harvard University Press, 1942.

Clement, D. et al. *Natural Semantics on the Computer.* Tech. rep. 416. INRIA, 1986.

Cocchiarella, N. B. 'Conceptual Realism as a Formal Ontology'. In: *Formal Ontology.* Ed. by R. Poli and P. Simons. Dordrecht: Kluwer, 1996, pp. 27–60.

Coffey, P. *The Science of Logic.* 2nd ed. Vol. 1. New York: Peter Smith, 1938.

Coquand, T. 'An algorithm for type-checking dependent types'. In: *Sci. Comput. Program.* 26.1-3 (1996), pp. 167–177.

Curry, H. B. 'Functionality in Combinatory Logic'. In: *Proc. of the National Academy of Sci. (U.S.)* 20.11 (1934), pp. 584–590.

— 'Remarks on the definition and nature of mathematics'. In: *Dialectica* 8 (1954), pp. 228–233.

— *Foundations of Mathematical Logic.* 2nd ed. Dover, 1977.

van Dalen, D. *Logic and Structure.* 3rd ed. Springer, 1997.

De Morgan, A. 'On the Structure of the Syllogism'. In: *Trans. Cambridge Philos. Soc.* 8 (1846), pp. 379–408.

— *Formal Logic.* London: Taylor and Walton, 1847.

Dedekind, R. 'Continuity and Irrational Numbers'. In: *Essays on the Theory of Numbers.* Trans. by W. W. Beman. New York: Dover, 1901.

— 'The Nature and Meaning of Numbers'. In: *Essays on the Theory of Numbers.* Trans. by W. W. Beman. New York: Dover, 1901.

Descartes, R. *Discours de la Méthode.* Leiden: J. Maire, 1637.

— *La géométrie.* Paris: A. Hermann, 1886.

— 'Rules for the Direction of the Mind'. In: *The Philosophical Writings of Descartes.* Trans. by D. Murdoch. Cambridge University Press, 1985.

— 'The World or Treatise on Light'. In: *The Philosophical Writings of Descartes.* Trans. by R. Stoothoff. Cambridge University Press, 1985.

Di Bella, S. 'Causa Sive Ratio. Univocity of Reason and Plurality of Causes in Leibniz'. In: *Leibniz : What Kind of Rationalist ?* Vol. 13. Logic, Epistemology, and the Unity of Science. Springer, 2008. Chap. 32.

Dijkstra, E. W. 'Under the spell of Leibniz's dream'. In: *Inf. Process. Lett.* 77.2–4 (2001), pp. 53–61.

Diogenes, L. *Lives of Eminent Philosophers.* Trans. by R. D. Hicks. 1 & 2 vols. Loeb Classical Library. Harvard University Press, 1925.

Diophantus. *Opera Omnia, cum Graecis Commentariis. Arithmeticorum Liber Sex et de Polygonis Numeris Liber.* Ed. by P. Tannery. Lipzig: B. G. Teubner, 1893.

Dirichlet, L. 'Sur la convergence des séries trigonométriques qui servent à représenter une fonction arbitraire entre des limites données'. In: *J. reine angew. Math.* 4 (1829), pp. 157–169.

Dybjer, P. 'Internal type theory'. In: *Types for Proofs and Programs.* Ed. by S. Berardi and M. Coppo. Vol. 1158. Lect. Notes Comput. Sc. Springer, 1996, pp. 120–134.

Empiricus, S. *Outlines of Pyrrhonism.* Trans. by R. G. Bury. Vol. 1. Loeb Classical Library. Harvard University Press, 1961.

Euclid. *Elements.* Ed. T. L. Heath. Santa Fe: Green Lion, 2002.

Euler, L. *Introductio in Analysin Infinitorum.* Lausanne: Marcus-Michael Bousquet & Co., 1748.

— *Institutiones Calculi Differentialis*. Saint Petersburg: Academia Imperialis Scientiarum, 1755.
— *Institutionum Calculi Integralis*. Vol. 1. Saint Petersburg: Academia Imperialis Scientiarum, 1768.
— *Elements of Algebra*. Trans. by J. Hewlett. St. Albans, England: Tarquin Publications, 2006.
Floyd, R. W. 'Assigning meaning to programs'. In: *Mathematical Aspects of Computer Science*. Providence: AMS, 1967, pp. 19–32.
Frege, G. *Begriffsschrift*. Halle: Louis Nebert, 1879.
— *Die Grundlagen der Arithmetik*. Breslau: W. Koebner, 1884.
— *Grundgesetze der Arithmetik I*. Vol. 1. Jena: Hermann Pohle, 1893.
— *Grundgesetze der Arithmetik II*. Vol. 2. Jena: Hermann Pohle, 1903.
— 'Function and Concept'. In: *Translations from the Philosophical Writings of Gottlob Frege*. Trans. by P. T. Geach. Oxford: B. Blackwell, 1960, pp. 21–41.
— 'On Concept and object'. In: *Translations from the Philosophical Writings of Gottlob Frege*. Trans. by P. T. Geach. Oxford: B. Blackwell, 1960, pp. 42–55.
Gadamer, H.-G. *Truth and Method*. Trans. by J. Weinsheimer and D. G. Marshall. 2nd ed. London & New York: Continuum, 1989.
Geach, P. T. 'The law of excluded middle'. In: *Supp. Proc. Arist. Soc.* 30 (1956), pp. 59–90.
— 'History of a Fallacy'. In: *J. Phil. Ass.* 5.19–20 (1958).
— 'Identity'. In: *Review of Metaphysics* 21.1 (1967), pp. 3–12.
— *Logic Matters*. Oxford: Blackwell, 1972.
Gentzen, G. 'Untersuchungen über das logische Schließen I & II'. In: *Math. Zeit.* 39 (1935), Pp. 176–210 & 405–431.
— 'Die Widerspruchfreiheit der reinen Zahlentheorie'. In: *Math. Ann.* 112 (1936), pp. 493–565.
Glivenko, V. 'Sur quelques points de la logique de M. Brouwer'. In: *Académie Royale de Belgique, Bulletin* 15 (1929), pp. 183–188.
Gödel, K. 'Über formal unentscheidbare Sätze der Principia Mathematica und verwandter Systeme I'. In: *Monatshefte für Mathematik und Physik* 38 (1931), pp. 173–198.
— 'Zur intuitionistische Arithmetik und Zahlentheorie'. In: *Ergebnisse eines mathematischen Kolloquiums* 4 (1933), pp. 34–38.
— 'Über die Lange von Beweisen'. In: *Ergebnisse eines mathematischen Kolloquiums* 7 (1936), pp. 23–24.
— 'Über eine bisher noch nicht benütze Erweiterung des finiten Standpunktes'. In: *Dialectica* 12.3 (1958), pp. 280–287.
Goodstein, R. L. 'Function Theory in an Axiom-Free Equation Calculus'. In: *Proc. London Math. Soc.* Ser. 2 48 (1945), pp. 401–434.
— 'The Arabic Numerals, Numbers and the Definition of Counting'. In: *Math. Gazette* 40.332 (1956), pp. 114–129.
— *Recursive number theory*. Amsterdam: North-Holland, 1957.
Gray, R. 'Georg Cantor and Transcendental Numbers'. In: *Amer. Math. Monthly* 101.9 (1994), pp. 819–832.

Gredt, I. *Elementa Philosophiae Aristotelico-Thomisticae*. 4th ed. Vol. 1. Freiburg: Herder & Co., 1925.

Hare, R. 'Meaning and Speech Acts'. In: *The Philosophical Review* 79.1 (1970), pp. 3–24.

van Heijenoort, J., ed. *From Frege to Gödel. A source book in mathematical logic, 1879–1931*. Cambridge Mass.: Harvard University Press, 1967.

Herbrand, J. 'Sur la non-contradiction de l'arithmétique'. In: *J. reine angew. Math.* 166 (1931), pp. 1–8.

Heyting, A. 'Die formalen Regeln der intuitionistischen Logik'. In: *Sitzungsberiche der Preussischen Akademie der Wissenschaften* (1930), pp. 42–56.

— 'Sur la logique intuitionniste'. In: *Académie Royale de Belgique, Bulletin* 16 (1930), pp. 957–963.

— *Intuitionism : An Introduction*. Amsterdam: North-Holland, 1956.

Hilbert, D. 'Über die Theorie der algebraischen Formen'. In: *Math. Ann.* 36.4 (1890), pp. 473–534.

— 'Mathematical problems'. Trans. by M. W. Newson. In: *Bull. Amer. Math. Soc.* 8.10 (1902), pp. 437–479.

— 'Über das Unendliche'. In: *Math. Ann.* 95.1 (1926), pp. 161–190.

— 'On the infinite'. In: *From Frege to Gödel. A source book in mathematical logic, 1879–1931*. Ed. by J. van Heijenoort. Cambridge Mass.: Harvard University Press, 1967, pp. 367–392.

Hilbert, D. and P. Bernays. *Grundlagen der Mathematik*. Vol. 1. Berlin: Springer, 1934.

Hofmann, M. 'Extensional concepts in intensional type theory'. PhD thesis. LFCS Edinburgh, 1995.

— 'Syntax and Semantics of Dependent Types'. In: *Semantics and Logics of Computation*. Ed. by A. M. Pitts and P. Dybjer. Vol. 14. Cambridge University Press, 1997, pp. 79–130.

Howard, W. A. 'The formulae-as-types notion of construction'. In: *To H. B. Curry : Essays on Combinatory Logic, Lambda Calculus and Formalism*. Ed. by J. P. Seldin and J. R. Hindley. Academic Press, 1980, pp. 479–490.

Husserl, E. *Logische Untersuchungen*. 3rd ed. Vol. 1. Halle: M. Niemeyer, 1922.

— *Logische Untersuchungen*. 5th ed. Vol. 2. Tübingen: M. Niemeyer, 1968.

— *The Crisis of European Sciences and Transcendental Phenomenology*. Trans. by D. Carr. Evanston: Northwestern University Press, 1970.

Ifrah, G. *Räknekonstens kulturhistoria*. Trans. by B. Ellenberger. Vol. 2. Original in French : *Histoire universelle des chiffres*. Wahlström & Widstrand, 2002.

Iverson, K. E. *A Programming Language*. Hoboken: John Wiley & Sons, 1962.

Jacobs, B. *Categorical Logic and Type Theory*. Studies in Logic and the Foundations of Mathematics 141. Amsterdam: North Holland, 1999.

Kant, I. *Kritik der reinen Vernunft*. Vol. 3. Kants Werke. Berlin: W. de Gruyter & Co., 1968.

Katz, J. J. and J. A. Fodor. 'The Structure of a Semantic Theory'. In: *Language* 39.2 (1963), pp. 170–210.

Kleene, S. C. 'Lambda-Definability and Recursiveness'. In: *Duke Math. J.* 2.2 (1936), pp. 340–353.

— 'On the Interpretation of Intuitionistic Number Theory'. In: *J. Symb. Log.* 10.4 (1945), pp. 109–124.

Klein, J. *Greek Mathematical Thought and the Origin of Algebra.* Trans. by E. Brann. New York: Dover, 1968.

Kline, M. *Mathematical Thought from Ancient to Modern Times.* Vol. 1. Oxford University Press, 1972.

Kneale, W. and M. Kneale. *The Development of Logic.* Oxford University Press, 1962.

Knuth, D. E. *The Art of Computer Programming.* Vol. 1. Addison-Wesley, 1968.

Kolmogorov, A. 'Zur Deutung der intuitionistischen Logik'. In: *Math. Zeit.* 35 (1932), pp. 58–65.

— 'On the principle of excluded middle'. In: *From Frege to Gödel. A source book in mathematical logic, 1879–1931.* (Original in Russian from 1925.) Cambridge Mass.: Harvard University Press, 1967, pp. 416–437.

Krivine, J.-L. *Introduction to Axiomatic Set Theory.* Dordrecht: Reidel, 1971.

Kronecker, L. 'Über den Zahlbegriff'. In: *J. reine angew. Math.* 101 (1889), pp. 337–355.

Landin, P. J. 'The mechanical evaluation of expressions'. In: *Comput. J.* 6.4 (1964), pp. 308–320.

— 'The next 700 programming languages'. In: *Communications of the ACM archive* 9.3 (1966), pp. 157–166.

Leibniz, G. W. *Dissertatio de Arte Combinatoria.* Leipzig: Fick & Seubold, 1666.

— 'Nova methodus pro maximis et minimis, itemque tangentibus, quae nec fractas nec irrationales quantitats moratur, et singulare pro illis calculi genus'. In: *Acta Eruditorum* (1684).

— 'Monadologia. Principia Philosophiæ'. In: *Acta Eruditorum* (1714).

— 'Projet et Essais pour arriver à quelque certitude pour finir une bonne partie des disputes et pour avancer l'art d'inventer'. In: *Opuscules et Fragments Inédits de Leibniz.* Ed. by L. Couturat. Paris: F. Alcan, 1903, pp. 175–182.

— *Philosophical Papers and Letters.* Ed. by L. E. Loemker. Dordrecht: Reidel, 1969.

— *Mathematische Schriften.* Vol. 5. Hildesheim: Georg Olms Verlag, 1971.

Leśniewski, S. 'Grundzüge eines neuen Systems der Grundlagen der Mathematik'. In: *Fundamenta Mathematicae* 14 (1929), pp. 1–81.

Locke, J. *An Essay Concerning Humane Understanding.* 2nd ed. London: Thomas Basset, 1694.

Maddy, P. 'Mathematical existence'. In: *Bull. Symb. Log.* 11.3 (2005), pp. 351–376.

Magnusson, L. 'The Implementation of ALF — A Proof Editor based on Martin-Löf's Monomorphic Type Theory with Explicit Substitution'. PhD thesis. Chalmers University of Technology, 1995.

Mancosu, P. *Philosophy of Mathematics and Mathematical Practice in the Seventeenth Century.* Oxford University Press, 1996.

— ed. *From Brouwer to Hilbert.* New York: Oxford University Press, 1998.

Marciszewski, W. 'The principle of comprehension as a present-day contribution to mathesis universalis'. In: *Philosophia Naturalis* 21 (1984), pp. 523–537.

Maritain, J. *The Degrees of Knowledge.* Trans. by R. McInerny. Notre Dame, Indiana: University of Notre Dame Press, 1995.

Markov, A. A. 'The theory of algorithms'. In: *Amer. Math. Soc. Transl.* Ser. 2 15 (1960), pp. 1–14.

Martin-Löf, P. 'An intuitionistic theory of types: predicative part'. In: *Logic Colloquium '73.* Ed. by H. E. Rose and J. Shepherdson. Amsterdam: North-Holland, 1975, pp. 73–118.

— 'About models for intuitionistic type theories and the notion of definitional equality'. In: *Proceedings of the Third Scandinavian Logic Symposium.* Vol. 82. Studies in Logic and the Foundation of Mathematics. North-Holland, 1975, pp. 81–109.

— 'Constructive mathematics and computer programming'. In: *Logic, Methodology and Philosophy of Science VI.* Ed. by L. J. Cohen et al. Amsterdam: North-Holland, 1982, pp. 153–175.

— *Intuitionistic Type Theory.* Studies in Proof Theory. Napoli: Bibliopolis, 1984.

— 'Truth of a proposition, evidence of a judgement, validity of a proof'. In: *Synthese* 73 (1987), pp. 407–420.

— 'A Path from Logic to Metaphysics'. In: *Atti del Congresso Nuovi Problemi della Logica e della Filosofia Scienza, Viareggio, 1990.* Vol. 2. Bologna: Clueb, 1991, pp. 141–149.

— 'Analytic and synthetic judgements in type theory'. In: *Kant and Contemporary Epistemology.* Ed. by P. Parrini. Kluwer, 1994, pp. 87–99.

— 'Verificationism Then and Now'. In: *The Foundational Debate; Complexity and Constructivity in Mathematics and Physics.* Ed. by W. De Pauli-Schimanovich, E. Köhler and F. Stadler. Kluwer, 1995, pp. 187–196.

— 'On the meanings of the logical constants and the justifications of the logical laws'. In: *Nordic Journal of Philosophical Logic* 1.1 (1996), pp. 11–60.

McLarty, C. 'Poincaré: Mathematics & Logic & Intuition'. In: *Phil. Math.* 5.2 (1997), pp. 97–115.

Michie, D. 'Memo Functions and Machine Learning'. In: *Nature* 218 (1968), pp. 19–22.

Mill, J. S. *A System of Logic.* New York: Harper & Brothers, 1846.

Milner, R., M. Tofte and R. Harper. *The definition of Standard ML.* Cambridge, U.S.: MIT Press, 1990.

Moore, G. E. 'The nature of judgement'. In: *Mind* 8.30 (1899), pp. 176–193.

— 'Certainty'. In: *Philosophical papers.* London: Allen and Unwin, 1959.

Moreau, L. 'A Syntactic Theory of Dynamic Binding'. In: *J. Higher-Order Sym. Comput.* 11.3 (1998), pp. 233–279.

Nordström, B., K. Petersson and J. M. Smith. *Programming in Martin-Löf's Type Theory.* Oxford University Press, 1990.

— 'Martin-Löf's Type Theory'. In: *Handbook of Logic in Computer Science.* Ed. by S. Abramsky, D. M. Gabbay and T. S. E. Maibaum. Vol. 5. Oxford University Press, 2000. Chap. 1, pp. 1–37.

Peano, G. *Arithmetices Principia Nova Methodo Exposita.* Turin: Fratelli Bocca, 1889.

Peirce, C. S. 'On the Algebra of Logic'. In: *Amer. J. Math.* 3.1 (1880), pp. 15–57.

— 'On the Algebra of Logic: a contribution to the philosophy of notation'. In: *Amer. J. Math.* 7.2 (1885), pp. 180–202.

Perlis, A. J. 'Epigrams on Programming'. In: *SIGPLAN Notices* 17.9 (1982), pp. 7–13.

Plotkin, G. D. *A Structural Approach to Operational Semantics.* Tech. rep. DAIMI FN-19. University of Aarhus, 1981.

Poincaré, H. *La Science et l'Hypothèse.* Paris: Flammarion, 1902.

— 'Les mathématiques et la logique'. In: *Revue de métaphysique et de morale* 14 (1906), pp. 294–317.

Poinsot, J. *The Material Logic of John of St. Thomas. Basic Treatises.* Trans. by Y. R. Simon, J. J. Glanville and G. D. Hollenhorst. Chicago: The University of Chicago Press, 1955.

Post, E. L. 'Finite Combinatory Processes – Formulation 1'. In: *J. Symb. Log.* 1 (1936), pp. 103–105.

— 'Formal Reductions of the General Combinatorial Decision Problem'. In: *Amer. J. Math.* 65 (1943), pp. 197–215.

Prawitz, D. *Natural Deduction. A Proof-Theoretical Study.* Stockholm: Almqvist & Wiksel, 1965.

Quine, W. V. 'A proof procedure for quantification theory'. In: *J. Symb. Log.* 20.2 (1955), pp. 141–149.

— *Theories and Things.* Cambridge: Harvard University Press, 1981.

— *Methods of Logic.* 4th ed. Cambridge: Harvard University Press, 1982.

Ramsey, F. P. 'Mathematical Logic'. In: *Math. Gazette* 13.184 (1926), pp. 185–194.

Ranta, A. *Type-Theoretical Grammar.* Oxford: Clarendon, 1995.

Recorde, R. *The Whetstone of Witte.* Amsterdam: Da Capo Press, 1969.

Richard, J. 'Les Principes des Mathématiques et le Problème des Ensembles'. In: *Revue Générale des Sciences Pures et Appliquées* 16.12 (1905), p. 541.

Russell, B. *The Principles of Mathematics.* Cambridge University Press, 1903.

— 'On Denoting'. In: *Mind* 14.56 (1905), pp. 479–493.

— 'On some difficulties in the theory of transfinite numbers and order types'. In: *Proc. London Math. Soc.* Ser. 2 4.1 (1906), pp. 29–53.

— 'The Theory of Implication'. In: *Amer. J. Math.* 28.2 (1906), pp. 159–202.

— 'Mathematical Logic as Based on the Theory of Types'. In: *Amer. J. Math.* 30.3 (1908), pp. 222–262.

Salvesen, A. *Polymorphism and monomorphism in Martin-Löf's Type Theory.* Tech. rep. Norwegian Computing Center, 1988.

Sato, M. et al. 'Calculi of Meta-variables'. In: *Lect. Notes Comput. Sc.* 2803 (2003), pp. 484–497.

Schütte, K. *Beweistheorie.* Berlin: Springer, 1960.

Scott, D. and C. Strachey. *Toward a Mathematical Semantics for Computer Languages.* Programming Research Group Technical Monograph PRG-6. Oxford Univ. Computing Lab., 1971.

Scotus, J. Duns. 'De rerum principio'. In: *Quaestiones Subtilissime Scoti in Metaphysicam Aristotelis. Eiusdem de Primo Rerum Principio Tractatus Atque Theoremata.* Ed. by M. Hibernicus. Venice: Bonetus Locatellus, 1497.

Searle, J. R. *Expression and Meaning.* Cambridge: Cambridge University Press, 1979.

Sebestik, J. 'Bolzano's Logic'. In: *The Stanford Encyclopedia of Philosophy.* Ed. by E. N. Zalta. 2007.

Setzer, A. 'Extending Martin-Löf's Type Theory by one Mahlo-universe'. In: *Arch. Math. Logic* 39 (2000), pp. 155–181.

Simpson, J. and E. Weiner, eds. *The Oxford English Dictionary.* 2nd ed. Oxford University Press, 1989.

Simpson, S. G. 'Logic and mathematics'. In: *The Examined Life, Readings from Western Philosophy from Plato to Kant.* Ed. by S. Rosen. New York: Random House, 2000, pp. 577–605.

Skolem, Th. A. 'Some remarks on axiomatized set theory'. In: *From Frege to Gödel. A source book in mathematical logic, 1879–1931.* Ed. by J. van Heijenoort. Cambridge Mass.: Harvard University Press, 1967, pp. 290–301.

Steele, G. L. *Common Lisp the Language.* 2nd ed. Digital Press, 1990.

Sundholm, G. 'Existence, Proof, and Truth-Making: A Perspective on the Intuitionistic Conception of Truth'. In: *Topoi* 13 (1994), pp. 117–126.

— 'Inference versus Consequence'. In: *The LOGICA Yearbook 1997.* Ed. by T. Childers. Filosofia Publishers, 1998, pp. 26–35.

Tasistro, A. 'Formulation of Martin-Löf's Type Theory with Explicit Substitutions'. Licentiate thesis. Chalmers University of Technology, 1993.

Troelstra, A. S. *Principles of Intuitionism.* Springer, 1969.

Turing, A. M. 'On Computable Numbers, with an Application to the Entscheidungsproblem'. In: *Proc. London Math. Soc.* 2.42 (1936), pp. 230–265.

— 'Systems of logic based on ordinals'. In: *Proc. London Math. Soc.* Ser. 2 45.2239 (1939), pp. 161–228.

Turner, W. 'Raymond Lully'. In: *The Catholic Encyclopedia.* Ed. by C. G. Herbermann et al. Vol. 12. New York: Robert Appleton Company, 1911.

Uspensky, V. A. and A. L. Semenov. 'What are the gains of the theory of algorithms: basic developments connected with the concept of algorithm and with its application in mathematics'. In: *Algorithms in Modern Mathematics and Computer Science.* Vol. 122. Lect. Notes Comput. Sc. Springer, 1981, pp. 100–234.

Venn, J. *Symbolic Logic.* London: Macmillan, 1881.

Vieta, F. *Artem Analyticam Isagoge seu Algebra Nova.* 2nd ed. Leiden: David Lopes, 1635.

Virgil. *Georgics.* Trans. by H. R. Fairclough. 2nd ed. Loeb Classical Library. Harvard University Press, 1999.

Weil, A. *The Apprenticeship of a Mathematician.* Trans. by J. Gage. Basel: Birkhäuser, 1992.

Weyl, H. 'Der circulus vitiosus in der heutigen Begründung der Analysis'. In: *Jahresber. d. Deutsch. Math.-Vereinig.* 28 (1919), pp. 85–92.

— 'Über die neue Grundlagenkrise der Mathematik'. In: *Math. Zeit.* 10 (1921), pp. 39–79.

— 'Comments on Hilbert's second lecture on the foundations of mathematics'. In: *From Frege to Gödel. A source book in mathematical logic, 1879–1931.* Ed. by J. van Heijenoort. Cambridge Mass.: Harvard University Press, 1967, pp. 480–484.

Whitehead, A. N. and B. Russell. *Principia Mathematica.* Vol. 1. Cambridge University Press, 1910.

Wilkins, J. *An Essay Toward a Real Character and a Philosophical Language.* London: Gellibrand & Martyn, 1668.

Wittgenstein, L. *Tractatus Logico-Philosophicus.* Trans. by C. K. Ogden and F. P. Ramsey. London: Routledge & Kegan Paul, 1922.

— *On Certainty.* Ed. by G. E. M. Anscombe and G. H. von Wright. Trans. by Denis Paul and G. E. M. Anscombe. Oxford: Basil Blackwell, 1969.

Youschkevitch, A. P. 'The concept of function up to the middle of the 19th century'. In: *Arch. Hist. Exact Sci.* 16.1 (1976), pp. 37–85.

Zermelo, E. 'Neuer Beweis für die Möglichkeit einer Wohlordnung'. In: *Math. Ann.* 65 (1908), pp. 107–128.

— 'Untersuchungen über die Grundlagen der Mengenlehre—I'. In: *Math. Ann.* 65 (1908), pp. 261–281.

Index of Proper Names

Index of Subjects

MIX
Papier aus verantwortungsvollen Quellen
Paper from responsible sources
FSC® C105338

FSC
www.fsc.org

If you have any concerns about our products,
you can contact us on
ProductSafety@springernature.com

In case Publisher is established outside the EU,
the EU authorized representative is:
Springer Nature Customer Service Center GmbH
Europaplatz 3, 69115 Heidelberg, Germany

Printed by Libri Plureos GmbH
in Hamburg, Germany